SpringerBriefs in Microbiology

SpringerBriefs in Microbiology present concise summaries of cutting-edge research and practical applications across a wide spectrum of fields. Featuring compact, authored volumes of 50 to 125 pages, the series covers a range of content from professional to academic. Typical topics might include:

- A timely report of state-of-the art analytical techniques
- A bridge between new research results published in journal articles and a contextual literature review
- A snapshot of a hot or emerging topic
- An in-depth case study or clinical example
- A presentation of core concepts that students must understand in order to make independent contributions
- Best practices or protocols to be followed
- A series of short case studies

SpringerBriefs in Microbiology showcase basic and translational research from a global author community. Briefs allow authors to present their ideas and readers to absorb them with minimal time investment, and will be published as part of Springer's eBook collection, with millions of users worldwide. In addition, Briefs will be available for individual print and electronic purchase.

Briefs are characterized by fast, global electronic dissemination, standard publishing contracts, standardized manuscript preparation and formatting guidelines, and expedited production schedules. We aim for publication 8-12 weeks after acceptance.

Moonisa Aslam Dervash • Abrar Yousuf
Mohammad Amin Bhat • Munir Ozturk

Soil Organisms

Deciphering the Life Beneath Our Feet

Moonisa Aslam Dervash
Punjab Agricultural University
Regional Research Station
Ballowal Saunkhri, Punjab, India

Mohammad Amin Bhat
Punjab Agricultural University
Regional Research Station
Ballowal Saunkhri, Punjab, India

Abrar Yousuf
Punjab Agricultural University
Regional Research Station
Ballowal Saunkhri, Punjab, India

Munir Ozturk
Faculty of Science
Ege University
Bornova, Izmir, Türkiye

ISSN 2191-5385   ISSN 2191-5393   (electronic)
SpringerBriefs in Microbiology
ISBN 978-3-031-66292-8   ISBN 978-3-031-66293-5   (eBook)
https://doi.org/10.1007/978-3-031-66293-5

This Springer imprint is published by the registered company Springer Nature Switzerland AG
The registered company address is: Gewerbestrasse 11, 6330 Cham, Switzerland

If disposing of this product, please recycle the paper.

Dedicated to "Our Parents, Teachers, and Alma Maters"

Foreword

This book, "Soil Organisms: *Deciphering the Life Beneath Our Feet*," offers an extensive exploration of soil biology, ecology, and its intricate interactions with human activities. The journey begins with an introduction to soil organisms, emphasizing their resilience and vulnerability to ecological stress. These organisms are crucial in maintaining soil structure, fertility, and ecosystem services, acting as bioindicators of soil health. Amid the challenges of feeding a growing population, the chapter underscores the negative impact of agrochemicals on soil biota and advocates for sustainable practices like biodynamic agriculture to preserve soil health and biodiversity. It also highlights the importance of natural networking systems and ecosystem cybernetics for sustainable ecology, calling for conservation strategies and biological soil remediation to protect and restore fertile soils.

The chapter 1 delves into the diversity of soil biota, from bacteria and fungi to earthworms and small mammals, each playing a vital role in maintaining soil health and ecosystem functioning. Chapter 2 stresses upon the critical ecosystem services provided by soil biodiversity, such as nutrient cycling and carbon sequestration, and the threats posed by pollution and intensive agriculture. Conservation efforts are deemed essential for protecting soil biodiversity and ensuring sustainable agriculture and food security.

The concept of soil stress ecology is then examined in Chapter 3, focusing on how various stressors affect soil ecosystems and their subsequent impact on plant and microbial communities. Understanding these dynamics is vital for enhancing soil health, agricultural productivity, and environmental sustainability, urging the maintenance of soil ecosystem balance amid natural and anthropogenic pressures. On the other hand, the agricultural industry, while crucial for global food production, presents significant challenges to soil biota. Chapter 4 discusses the dual impact of agriculture on soil ecosystems, highlighting the need for integrated approaches that balance productivity with sustainability, conservation, and social equity.

Bioindicators are explored as essential tools for monitoring soil pollution, with plants, soil microorganisms, earthworms, and other indicators providing sensitive responses to changes in soil quality. The importance of these bioindicators in maintaining soil health is emphasized in Chapter 5, offering valuable insights for detecting and mitigating soil pollution. The book also addresses the role of microbes in organic farming in Chapter 6, highlighting how traditional practices and sustainable strategies can combat the detrimental effects of modern agriculture. The emphasis is on fostering soil health and productivity through microbial activity, paving the way for a more sustainable and regenerative food system.

The concept of the "fungal internet" reveals the intricate networks of mycorrhizal fungi that connect trees and facilitate communication and resource exchange. Chapter 7 underscores the importance of these natural networking systems in maintaining ecosystem health and resilience, with practical insights into their functions and benefits. Biodynamic agriculture is presented as a holistic approach that views the farm as a self-sustaining ecosystem, promoting environmental sustainability and high biodiversity. Chapter 8 discusses the principles, methods, and benefits of biodynamic agriculture, as well as its challenges and future prospects.

The field of ecosystem cybernetics is explored, detailing how soil organisms contribute to nutrient cycling, energy flow, and feedback mechanisms within ecosystems. The interconnectedness and emergent properties of ecosystems are emphasized in Chapter 9, along with the challenges and future directions of this scientific discipline. Finally, the book discusses soil remediation, particularly biological approaches like bioremediation, which utilize living organisms to restore contaminated soil. The role of hyperaccumulator plants, microorganisms, and fungi in mitigating pollution and enhancing soil health is highlighted. Chapter 10 also examines regulatory frameworks and the concept of a circular economy in addressing soil pollution and promoting sustainable land management practices.

In summary, "Soil Organisms: *Deciphering the Life Beneath Our Feet*" provides a profound understanding of soil ecosystems, their critical roles, and challenges and opportunities in soil management. It calls for a shift action toward more sustainable and regenerative practices to ensure the health and productivity of our soil for future generations. The authors have invested significant effort and dedication into crafting this comprehensive and novel volume on soil organisms. By meticulously researching and compiling the latest scientific insights, they have illuminated the fascinating and complex world beneath our feet. Their work not only highlights the critical roles soil organisms play in maintaining ecosystem health but also addresses the pressing challenges and opportunities for sustainable soil management. Through detailed analysis and thoughtful synthesis, this book offers a valuable resource for students, scientists, educators, and policymakers, fostering a deeper understanding and appreciation of the vital life forms that underpin our planet's ecological balance.

Former Professor, Centre of Research for Development   G. A. Bhat
University of Kashmir,
Srinagar, Jammu and Kashmir, India

Preface

Soil biota is an important and fundamental part of terrestrial ecology. Soil organisms include all those life forms that pass a significant proportion of their life within a soil profile. The range of organisms may vary from microscopic entities to macro levels. These organisms are the hidden beauties that take up the tasks of food chain regulation, organic matter decomposition, and nutrient enrichment in soil. Soil organisms can be grouped into three classes: chemical engineers (bacteria, fungi, and protozoans), biological regulators (small soil invertebrates like nematodes and mesofauna), and ecosystem engineers (large soil invertebrates and small mammals). These creatures are mainly regulated by certain physical and biological factors. The physicochemical characteristics of soil can determine its biological diversity, which influences the regulatory services in an ecosystem.

There are innumerable economic values of soil biodiversity that can be quantified in terms of maintaining the soil structure and fertility, carbon flux and climate change alleviation, water cycle regulation, decontamination through bioremediation, and pest control. Soil organisms are part and parcel of soil ecosystems with each organism having its own respective niche that weaves the warps and wefts of biological world, pulling out a single creature from the soil environment, will drastically impact the delicate organization of food web.

In the present era of climate change, population explosion, and other environmental problems, there is tremendous pressure on these soil creatures. Soil pollution and environmental upheavals can result in inflation of the services extended by soils and, moreover, can result in the spread of soil-borne diseases to humans. Soil degradation, altered land uses, climate change, anthropogenic soil pollution, impetus given to genetically modified crops, and invasive species can have direct and indirect consequences on soil biology. In order to address the threats with potential solutions, it is mandatory to understand life beneath our feet by comprehending the biological dynamics and services.

The aim of this brief book is to showcase the mechanisms of biological dynamics of soil organisms at micro, meso, and macro scales. The introductory chapters of this book focus on understanding the faunistic diversity and its significance in soil, bioindicators in the assessment of soil quality, and consequences of various anthropogenic stressors on soil, highlighting the eves and odds associated with agricultural industry vis-à-vis environmental pollution and possible influences on soil biota.

The succeeding chapters focus on the fungal internet, ecosystem cybernetics, and potential remedial technologies in terms of microbial organic farming and biodynamic agriculture for soil conservation, and the last chapter will focus on novel remediation strategies to deal with degraded soils.

This brief book will serve as a logical support for investigators with sufficient references for academicians, research scholars, and students. Researchers and connoisseurs shall find it a comprehensive manuscript regarding soil organisms, fulfilling the diverse needs of readers, staff in ministries, teaching staff, and researchers in this field.

Ballowal Saunkhri, Punjab, India   Moonisa Aslam Dervash
                                                                                Abrar Yousuf
                                                                    Mohammad Amin Bhat
Bornova, Izmir, Türkiye   Munir Ozturk

Acknowledgments

This manuscript has been written as a collective effort by the authors mentioned in this book. The authors are specially thankful to Dr. Kenneth Teng and his associate Mr. Arun Siva Shanmugam for support at every and each step. Our personal gratitude to the Springer Team for their patience and constructive collaboration.

Contents

About the Authors

Moonisa Aslam Dervash, PhD, Post Doc   has obtained her PhD from Division of Environmental Sciences, SKUAST-K, India. Her specialization is in environmental monitoring, environmental awareness, ecology, soil biology, wetland restoration, and carbon sequestration. She has authored 7 books with international publishers and has published more than 60 scientific research articles/book chapters in journals/books of international reputation. She is the recipient of many prestigious awards and felicitations for her dedicated accomplishments. Her focus has remained on many facets of society, especially on environmental conservation and women empowerment. She has been felicitated by the state government (J&K Department of Ecology. Environment and Remote Sensing) for her outstanding contribution to mass environmental awareness and conservation through electronic media. So far, she has extended nine years of dedicated service toward environmental awareness and conservation through a weekly radio program "Soun Aoundh Pukk" on All India Radio (2015–till date). She was recently awarded Postdoctoral Fellowship by Ministry of Education (ICSSR), Government of India (2022–2024). Currently, she is working in a project on "Identification of Potential Rainwater Harvesting Sites Using Geospatial Technologies in Kandi Region of Punjab" funded by Science and Engineering Research Board (SERB) at Punjab Agricultural University–Regional Research Station, Ballowal Saunkhri, SBS Nagar, Punjab.

Abrar Yousuf pursued his PhD from the Department of Soil and Water Engineering, Punjab Agricultural University, Ludhiana. Presently, he is working as a scientist (Soil and Water Engineering) at Punjab Agricultural University–Regional Research Station, Ballowal Saunkhri, SBS Nagar, Punjab. His field of specialization is watershed hydrology, soil erosion modeling, watershed management, remote sensing and GIS, rainwater management, and dryland agriculture. He is involved in continuous monitoring of runoff and sediment yield from various watersheds located in the Kandi region of Punjab. He has been involved in several research projects funded by various funding institutes, such as ICAR New Delhi, DST New Delhi, GIZ New Delhi, IPRO Consult Germany, Department of Soil and Water Conservation, Govt. of Punjab and SERB, New Delhi. He is working on the ex-situ management of rainwater in farm ponds and its judicious use through micro-irrigation systems. He has constructed farm ponds in various adopted villages. Dr. Abrar has authored a number of research articles in journals of national and international repute, technical bulletins, and two books. He is a member of several scientific associations and actively participates in scientific conferences, workshops, summer and winter schools, and training programs for building up a high scientific temper and technical skills. Recently, he has been felicitated with the "Best Thesis Award" for his doctoral research work by the Soil Conservation Society of India, New Delhi.

Mohammad Amin Bhat, PhD received his BSc (Agriculture) degree from SKUAST-Kashmir and MSc and PhD in Soil Science from CCS Haryana Agricultural University, Hisar. He was awarded ICAR-Junior Research Fellowship during the Master's program and a fellowship from the University Grants Commission, New Delhi, during the doctoral program. Dr. Bhat is working as a soil scientist at Regional Research Station, Punjab Agricultural University, Ballowal Saunkhri, Punjab. He has authored/coauthored more than 30 articles in refereed journals of national and international repute and several technical bulletins, book chapters, and popular articles. His field of specialization is soil chemistry and soil fertility. Presently, he is working on soil fertility management in dryland agroecosystems of Northwest India. He is a member of various scientific societies and an active participant in conferences, seminars, workshops, and summer and winter schools. Moreover, he is involved in teaching soil science to undergraduate students at PAU-College of Agriculture, Ballowal Saunkhri, Punjab.

Munir Ozturk, PhD, DSc received his BSc (Biology-Chemistry) degree from Sri Partap College, Kashmir; MSc from Postgraduate Department of Botany, Jammu and Kashmir University at Hazratbal, Kashmir; PhD and DSc from Ege University, Izmir, Turkiye. He has served at the Ege University-Turkiye for more than 50 years in different positions and has been founder director of the Centre for Environmental Studies, Ege University, and Chairman of the Botany Department as well as Director of the Botanical Garden. *Sideritis ozturkii* and *Verbascum ozturkii* are two newly recorded endemic plant species from Turkiye in his name. His fields of scientific interest are pollution and biomonitoring, plant ecophysiology, biosaline agriculture, and medicinal aromatic plant conservation. He has published almost 60 books with international publishers and more than 80 book chapters and 200 papers in international journals (120 with an impact factor); has presented 125 papers at international and 85 at national meetings. He has served as a guest editor for more than 13 international journals and holds many memberships of "institutions and professional bodies." He has received fellowships from the globally recognized Alexander von Humboldt Foundation, Japanese Society for Promotion of Science, and National Science Foundation of the USA. He has worked as consultant fellow at the Faculty of Forestry, Universiti Putra Malaysia, Malaysia; as distinguished visiting scientist at International Centre for Chemical and Biological Sciences, ICCBS-TWAS, Karachi University, Pakistan; and as "Vice President of the Islamic World Academy of Sciences" from 2017 to 2022. He is a "Fellow of the Islamic World Academy of Science" as well as "Foreign Fellow of Pakistan Academy of Science."

Chapter 1
Introduction to Soil Organisms:
Deciphering the Life Beneath Our Feet

Abstract Soil organisms exhibit varied resilience and susceptibility to ecological stress, necessitating comprehensive study and classification based on spatial and temporal factors. Their role in maintaining soil structure, fertility, and ecosystem services underscores the need for detailed understanding, especially amid environmental disturbances. Soil organisms also serve the role of bioindicators providing a coherent information regarding soil health and ecosystem functioning which needs to be thoroughly explored. In the present era of population explosion, conventional agriculture has emerged as an essential strategy for feeding a growing population, but agrochemical use often harms soil biota. Emphasizing sustainable practices like biodynamic agriculture can mitigate these effects, preserving soil health and biodiversity. Natural networking systems and ecosystem cybernetics, focusing on nutrient cycling and feedback mechanisms, are crucial for sustainable ecology. Conservation strategies and biological soil remediation are vital for protecting and restoring fertile soils, guiding future research and sustainable development efforts, as described in this chapter.

Soil biota is the natural assemblage of organisms which spend considerable portion of their lifespan in soil. Soils harbor huge depository of organic as well as mineral wealth, vital for life to flourish on Earth. Soil organisms play a significant role in decomposition of organic matter, regulation of nutrient and mineral cycles, facilitation of carbon fluxes, and maintaining soil fertility and health, backbone to boost agricultural and horticulture enterprises. Since soil biota is the treasure of diverse biological entities representing 25% of the total biodiversity on Earth (UNEP and FAO 2020), thus it is hectic to classify. Thus, on the basis of body dimensions, the soil organisms have been divided into four broader classes, viz., microorganisms (microscopic), mesofauna (between 0.2 and 2 mm), macrofauna (between 2 and 20 cm), and megafauna (greater than 20 cm). Each member occupies a specific niche in an ecosystem in an ecological setup (Briones 2014).

Different organisms possess different resilience as well as susceptible characteristics when exposed to ecological stress conditions. Soil ecology has been studied from time to time in different parts of the world, but there are still considerable lacunae which need to be showcased. The identification of soil biota on the basis of spatiotemporal aspects and their susceptibilities to global environmental upheavals is a need of the hour. The magnification of ecological services carried out by hidden beauties (microbial biomass) harboring the soil environment is to be understood at the basic level. Mesofauna and macrofauna are known to create small channels in soil which is important for maintaining porosity of soils, maintaining the water-holding capacities, and boosting the fertility of soils. The discourse on soil organisms is an important dialogue in magnifying the associations and linkages between various types of organism, ecosystem cybernetics, and myriad services offered by an ecosystem. In the present era of escalated environmental disturbances, soil quality assessment and management through soil organisms have emerged as an elusive concept (Ozturk et al. 2022). Therefore, the role of biological indicators is an important aspect in soil quality assessment and anticipatory measures of conservation based on interaction of fauna with soil under the prevailing environmental factors. Understanding soil organisms under stress conditions will aid us to bring forth the averments regarding the impacts of various environmental changes on soil biota.

In the current era of technological advancements and scientific achievements, life expectancy has escalated, whereas infant mortality rate has remarkably declined due to better medical advancements. It has altogether resulted in population explosion. Thus, in order to supply food and surplus commodities to sustain huge global population, agriculture is an unchallengeable warrior. In order to generate more food, we often ignore the irreversible harm that conventional agriculture poses on the environment. The detrimental effects of agrochemicals (pesticides and synthetic fertilizers) on various domains of environment eventually prove to be detrimental to food chain and soil organisms. The evens and odds associated with agricultural industry must be brought into limelight to understand the fragile technicalities of ecological dynamics. The ecosystem is a self-regulating system in terms of energy flow and material cycling, but under the negative or positive inputs, ecosystems behave as information networks. Some portion of the output further becomes input (in the form of feedback) for the system to continue the dynamics in an ecosystem, and in due process, a feedback which circulates in an ecosystem can be positive or negative. These information networks/feedback mechanisms under the influence of certain system inputs are based on flow of information which is an inherent characteristic of ecosystem. This mechanism of control and communication in an ecological system is known as cybernetics. Cybernetic nature of ecosystem can be understood through nutrient cycling, energy flow, feedback mechanisms, soil structure, and stability. In order to understand the impact of human-induced disturbances in an ecosystem, the role of microbes and cybernetics can play an outstanding role in terms of sustainable ecology.

On the other hand, the goals of sustainable development advocate the use of eco-friendly products with comparably less ecological footprints. Thus, encouragement of biodynamic agriculture shall be an important investment for sustainable development. The various strategies of biodynamic agriculture coupled with technological advancements can be very beneficial as far as climate-resilient sustainable agriculture is concerned. Besides, biodynamic agriculture or sustainable agriculture doesn't pose any detrimental impacts on soil fauna and soil health. To continue the pace of development under the banner of sustainable mode, it is imperative to conserve natural resources. Soil is the most important natural asset, and various biological and engineering strategies are available as far as conservation of soil is concerned. By conserving a soil environment, soil biodiversity can be conserved which is important in maintaining flows and fluxes in an ecosystem. On the order hand, today, fertile soils are endangered to various degrees due to intensive agricultural practices because the land resources are limited and the feeding mouths are escalating exponentially (Ozturk et al. 2021). Thus, it is essential to discover the strategies focused to remediate the polluted soils. The existing data on soil remediation through biological approaches would pave a way for future research priorities and potential outputs in scientific field.

Academicians, scholars, and students shall find it as a comprehensive manuscript regarding soil organisms and shall fulfill the needs of teaching staff and research scholars.

References

Briones MJI (2014) Soil fauna and soil functions: a jigsaw puzzle. Front Environ Sci 2:1–22

Ozturk M, Altay V, Efe R (2021) Biodiversity of West Asia-Caucasus: prospects and challenges for conservation and sustainable use, vol I. Springer Nature, p 655

Ozturk M, Khan S, Egamberdieva D, Khassanov F, Efe R, Altay V (2022) Biodiversity, conservation and sustainability in Asia—volume 2: prospects and challenges in south and middle Asia. Springer Nature, p 1070

UNEP (United Nations Environment Programme) and FAO (Food and Agriculture Organization of the United Nations) (2020) United Nations Ecosystem Restoration 2021–2030. Accessed at https://www.decadeonrestoration.org/

Chapter 2
Unveiling and Understanding the Soil Biota

Abstract Soil biota encompasses a diverse array of lifeforms, including algae, bacteria, fungi, nematodes, protozoa, arthropods, earthworms, and small mammals, which play fundamental roles in maintaining soil health, fertility, and ecosystem functioning. Soil biodiversity provides critical ecosystem services such as nutrient cycling, soil formation, water filtration, carbon sequestration, and pest regulation. Thus, it is imperative to explore each level of soil biodiversity to have a better understanding regarding their niches. However, soil biodiversity is under severe threats from anthropogenic activities like pollution and intensive agriculture. Conservation efforts are vital for protecting soil biodiversity and ensuring sustainable agriculture, food security, and environmental resilience, as discussed in this chapter.

2.1 Introduction

Soil biota (or soil biodiversity) refers to the diversity and abundance of life-forms dwelling in the soil ecosystem. It encompasses a wide range of organisms, including algae, bacteria, fungi, nematodes, protozoa, arthropods, earthworms, and small mammals (Ozturk et al. 2022). Soil biodiversity plays a fundamental role in maintaining soil health, fertility, and ecosystem functioning (Neher and Barbercheck 2019). However, soil biodiversity is under severe threats from anthropogenic activities like pollution and intensive agriculture.

Some of the key aspects of soil biodiversity include the following:

2.1.1 Microbial Diversity

Bacteria and fungi are the most abundant and diverse groups of soil organisms. They play critical roles in nutrient cycling, decomposition of organic matter, soil structure formation, and disease suppression (Abawi and Widmer 2000). The diversity of microbial communities in soil contributes to various ecosystem services, such as plant productivity and resilience to environmental stress.

2.1.2   Invertebrate Diversity

Soil invertebrates, including nematodes, mites, springtails, beetles, ants, and earthworms, are essential components of soil biodiversity. They contribute to nutrient cycling, decomposition, soil aeration, and pest regulation (Zeng et al. 2024). Invertebrates also influence soil structure and fertility through their burrowing activities and feeding behaviors.

2.1.3   Plant-Soil Feedbacks

Soil biodiversity interacts with plant communities through complex feedback mechanisms. Plants release root exudates and organic matter into the soil, shaping the composition and activity of soil microbial communities. In turn, soil microbes influence plant nutrient uptake, growth, and resistance to pathogens and pests (Molefe et al. 2023).

2.1.4   Ecosystem Services

Soil biodiversity provides numerous ecosystem services that are essential for human well-being and ecosystem sustainability (Wall and Nielsen 2012; Robb 2024). These services include nutrient cycling, soil formation, water filtration, carbon sequestration, pest regulation, and support for biodiversity aboveground.

2.1.5   Threats and Conservation

Soil biodiversity faces threats from various anthropogenic activities, including land-use alteration, habitat destruction, pollution, intensive agriculture, and climate change (Tibbett et al. 2020). Loss of soil biota can lead to degradation of soil quality, reduced agricultural productivity, and loss of ecosystem resilience. Conservation efforts to protect and enhance soil biodiversity include sustainable land management practices, organic farming, agroforestry, reforestation, and habitat restoration.

Overall, soil biodiversity is a vital component of terrestrial ecosystems, supporting ecosystem functions and services that are indispensable for sustainable agriculture, food security, and environmental conservation. Preserving and enhancing soil biodiversity are crucial for maintaining soil health, productivity, and resilience in the face of global environmental challenges, as discussed in this chapter.

2.2   Soil and Bacteria Friendship

Bacteria are one of the most abundant and diverse groups of soil organisms. They are involved in nutrient cycling, such as nitrogen fixation, decomposition of organic matter, and nutrient mineralization (Lladó et al. 2017). There are numerous types of bacteria found in soil, each with its own ecological niche and role in soil health and fertility. Some common types of beneficial soil bacteria are as under:

2.2.1   Nitrogen-Fixing Bacteria

These bacteria possess the capability to transform atmospheric nitrogen (N_2) into ammonia (NH_3) through a process called nitrogen fixation (Lindström and Mousavi 2020). Examples include species of *Rhizobium*, *Bradyrhizobium*, *Azotobacter*, and *Azospirillum*. Nitrogen-fixing bacteria form symbiotic relationships with certain plants, such as legumes, and provide them with a source of fixed nitrogen.

2.2.2   Nitrifying Bacteria

Nitrifying bacteria convert ammonium (NH^{4+}) into nitrites (NO^{2-}) and then into nitrates (NO^{3-}). This process, known as nitrification, makes nitrogen available for plant uptake (Grzyb et al. 2021). Common nitrifying bacteria include *Nitrosomonas* and *Nitrobacter*.

2.2.3   Denitrifying Bacteria

Denitrifying bacteria carry out the process of denitrification, in which nitrates (NO^{3-}) are converted back into nitrogen gas (N_2), which is then released into the atmosphere (Grzyb et al. 2021). This process contributes to nitrogen loss from the soil. Examples of denitrifying bacteria include *Pseudomonas* and *Bacillus* species.

2.2.4   Ammonifying Bacteria

Ammonifying bacteria decompose organic nitrogen compounds, such as proteins and amino acids, into ammonium (NH_4^+), which can then be utilized by plants or further transformed by other soil organisms (Lamb et al. 2014). Various types of bacteria, such as *Clostridium* and *Bacillus* species, are involved in ammonification.

2.2.5   Cellulolytic Bacteria

Cellulolytic bacteria break down cellulose, a complex carbohydrate found in plant cell walls, into simpler sugars. This process is important for the decomposition of plant litter and organic matter in soil (Datta 2024). Examples include species of *Cellulomonas*, *Bacillus*, and *Clostridium*.

2.2.6   Phosphate-Solubilizing Bacteria

These bacteria have the capacity to solubilize insoluble forms of phosphorus in the soil, making it available for plant uptake. Phosphate-solubilizing bacteria play a vital role in phosphorus cycling and plant nutrition (Tian et al. 2021). Examples include species of *Bacillus*, *Pseudomonas*, and *Rhizobium*.

2.2.7   Sulfur-Oxidizing and Sulfur-Reducing Bacteria

These bacteria are involved in the cycling of sulfur in the soil ecosystem. Sulfur-oxidizing bacteria oxidize elemental sulfur or sulfide compounds into sulfate, while sulfur-reducing bacteria reduce sulfate back into sulfide (Gonsior et al. 2018). *Thiobacillus* and *Desulfovibrio* are examples of sulfur-oxidizing and sulfur-reducing bacteria, respectively.

These are just a few examples of the diverse array of bacteria found in soil ecosystems. Each type of bacteria plays a crucial character in nutrient cycling, decomposition, and overall soil fertility.

2.3   Soil Fungi: Types and Role

Fungi are important decomposers in the soil ecosystem. They break down complex organic compounds into simpler forms, releasing nutrients that are then available to plants (Bahram and Netherway 2022).

There are several types of beneficial soil fungi, many of which form symbiotic relationships with plants and contribute to soil health and plant growth. Few common types are as follows:

2.3.1 Arbuscular Mycorrhizal Fungi (AMF)

AMF form symbiotic associations with the roots of the majority of plant species, including many agricultural crops, grasses, and wild plants. They penetrate the root cells and form highly branched structures called arbuscules and vesicles, increasing the surface area available for nutrient exchange between the fungus and the plant. AMF are particularly important for enhancing phosphorus uptake by plants and can also improve their resistance to certain stresses, such as drought and disease (Ebbisa 2023).

2.3.2 Ectomycorrhizal Fungi

Ectomycorrhizal fungi form associations primarily with trees, especially in temperate and boreal forests (Policelli et al. 2020). Unlike AMF, they do not enter the plant cells but instead form a dense network of hyphae around the outside of the plant roots and between root cells. Ectomycorrhizal fungi are important for enhancing nutrient uptake, particularly nitrogen and phosphorus, and they can also improve resistance to pathogens and environmental stresses.

2.3.3 Trichoderma spp

Trichoderma species are commonly present in soil and are known for their capability to promote plant growth and protect against soilborne pathogens. They produce enzymes that degrade organic matter and reduce the growth of harmful fungi through competition and parasitism (Yao et al. 2023). *Trichoderma* spp. are used in biological control strategies to manage plant diseases and improve soil heal.

2.3.4 Mycofungicides

Some fungi have been developed into commercial mycofungicides for controlling soilborne pathogens. Examples include species of *Trichoderma*, *Beauveria*, *Metarhizium*, and *Gliocladium*. These beneficial fungi can be applied to soil or plant roots to suppress pathogens and enhance plant growth while decreasing the requirement for synthetic chemical pesticides (Yao et al. 2023).

2.3.5   Endophytic Fungi

These fungi inhabit the internal tissues of plants without causing disease symptoms. Many endophytic fungi provide benefits to their host plants, such as improved nutrient uptake, better stress tolerance, and protection against pathogens and pests (Gowtham et al. 2024). They can also promote soil health indirectly by persuading the quality and quantity of plant litter and root exudates.

These are just a few examples of the diverse array of beneficial soil fungi. They play essential roles in nutrient cycling, plant health, and ecosystem resilience, highlighting the importance of maintaining and enhancing soil fungal diversity for sustainable agriculture and environmental conservation.

2.4   Algae in Soil

Algae are photosynthetic organisms commonly associated with aquatic environments, but they can also be found in soil, especially in moist or wet habitats. While algae are less abundant in soil compared to other microorganisms like bacteria and fungi, they can still play important roles in soil ecosystems (Crouzet et al. 2019). The magnified look at soil algae is discussed as under:

2.4.1   Types of Algae

Several types of algae may be present in soil, such as cyanobacteria, green algae, diatoms, and other microalgae. These organisms vary in size, shape, and pigmentation.

2.4.2   Habitats

Algae in soil are often found in environments with sufficient moisture, such as around bodies of water, in wetlands, or in areas with high organic matter content. They may also inhabit the upper layers of soil where moisture levels are higher.

2.4.3   Contribution to Soil Health

Algae contribute to soil health and fertility in several ways:

(i) *Photosynthesis*: Like algae in aquatic environments, soil algae conduct photo-synthesis, producing organic matter and oxygen (Jassey et al. 2022). This pro-cess contributes to carbon sequestration and soil organic carbon content.

(ii) *Nitrogen Fixation*: Some species of cyanobacteria are capable of biological nitrogen fixation (Battistuzzi et al. 2023). This process helps enrich the soil with nitrogen, an essential nutrient for plant growth.

(iii) *Soil Stability*: Algae can help stabilize soil particles and prevent erosion by forming a network of filaments or mats that bind soil particles together (Crouzet et al. 2019). This is particularly important in areas prone to erosion, such as riverbanks or disturbed soils.

(iv) *Microbial Interactions*: Algae may interact with other soil microbes, such as bacteria and fungi, influencing nutrient cycling and regulation of microbial communities (Solomon et al. 2023).

2.4.4   Challenges and Considerations

While algae can provide benefits to soil ecosystems, excessive algal growth may also indicate nutrient imbalances or environmental disturbances. In some cases, algal blooms in soil may lead to the depletion of oxygen levels or the release of toxins harmful to other organisms (Anabtawi et al. 2024).

Overall, while algae are not as prominent in soil ecosystems as bacteria and fungi, they still play important roles in ecosystem functioning, nutrient cycling, and soil health. Understanding the dynamics of algae in soil environments can help inform soil management practices and promote sustainable land use.

2.5   Protozoa in Soil

Protozoa are single-celled organisms that belong to the kingdom Protista. They are abundant in soil ecosystems and play important roles in decomposition, nutrient cycling, and regulation of microbial populations (Geisen et al. 2018). The detailed account of protozoa in soil is as follows:

2.5.1   Types of Protozoa

Soil protozoa encompass a diverse range of taxa, including amoebae, flagellates, cili-ates, and testate amoebae. These organisms vary in size, shape, and feeding strategies.

2.5.2   Feeding Strategies

Protozoa in soil are classified into different groups based on their feeding habits:

 (i) *Bacterivores*: Bacterivorous protozoa feed primarily on bacteria, grazing on microbial populations and regulating bacterial abundance in soil.
 (ii) *Fungivores*: In soil, fungivorous protozoa consume fungi, contributing to the decomposition of fungal biomass and nutrient cycling.
(iii) *Herbivores:* Some protozoa feed on algae, plant debris, or other organic matter present in soil.
(iv) *Omnivores*: Omnivorous protozoa have a broader diet, consuming a variety of prey including bacteria, fungi, and other protozoa.

2.5.3   Nutrient Cycling

Protozoa play strategic roles in nutrient cycling and mineralization in soil ecosystems. By consuming microbial biomass, protozoa release nitrogen, phosphorus, and carbon in forms that are readily available to plants and other organisms (Geisen et al. 2018).

2.5.4   Predation and Control of Microbial Populations

Protozoa are important predators of bacteria and fungi in soil. They help regulate microbial populations, preventing the overgrowth of certain species and maintaining microbial diversity.

2.5.5   Interactions with Plants and Soil Organisms

Protozoa interact with other soil organisms, including plants, bacteria, fungi, and nematodes. For example, some protozoa form symbiotic affiliations with plant roots, contributing to nutrient uptake and plant growth (Chamkhi et al. 2022). Protozoa may also serve as prey for soil fauna, such as nematodes or predatory mites.

2.5.6   Environmental Factors

The presence and functions of soil protozoa are influenced by environmental factors such as temperature, pH, soil moisture, and nutrient availability. Protozoa are often dominant in soils with higher organic matter content and in habitats with moderate moisture levels.

Overall, protozoa are integral components of soil ecosystems, contributing to decomposition, nutrient cycling, and regulation of microbial communities. The role of protozoa in soil ecology is important to be understood for sustaining soil health and ecosystem functioning in agricultural as well as in natural environments.

2.6 Nematodes in Soil

Nematodes are a diverse group of microscopic roundworms that inhabit soil ecosystems. They play various roles in soil ecology, contributing to nutrient cycling, decomposition, and regulation of microbial populations (Khanum et al. 2022). The detailed account of soil nematodes is as under:

2.6.1 Diversity

Nematodes are one of the most diverse groups of soil fauna, with thousands of species found in soil ecosystems worldwide. They vary in size, shape, feeding habits, and ecological functions.

2.6.2 Feeding Strategies

Based on their feeding habits, nematodes in soil can be classified into the following:

2.6.3 Bacterivores

Bacterivorous nematodes feed primarily on bacteria, grazing on microbial populations and regulating bacterial abundance in soil.

2.6.4 Fungivores

Fungivorous nematodes consume fungi, contributing to the decomposition of fungal biomass and nutrient cycling in soil.

2.6.5 Predators

Predatory nematodes feed on other soil organisms, including bacteria, fungi, other nematodes, and even small arthropods. They help to control populations of other soil organisms and contribute to the regulation of soil food webs.

2.6.6 Plant Parasites

Some nematodes are plant parasites, feeding on plant roots and causing damage to crops. Plant-parasitic nematodes can have significant economic impacts on agriculture by reducing crop yields and causing plant diseases.

2.6.7 Nutrient Cycling

Nematodes play imperative functions in nutrient cycling and mineralization in soil ecosystems. By consuming microbial biomass and organic matter, nematodes release nitrogen, phosphorus, and carbon in forms that are readily available to plants and other organisms (Li et al. 2018).

2.6.8 Interactions with Plants

Nematodes interact with plants in various ways. Some nematodes form mutualistic relationships with plants, such as mycorrhizal nematodes that associate with mycorrhizal fungi and augment nutrient uptake by plants. However, plant-parasitic nematodes can cause damage to plant roots, leading to reduced plant growth and productivity (Li et al. 2018).

2.6.9 Environmental Factors

The abundance and activity of nematodes in soil are subjective to environmental factors, for example, temperature, pH, soil moisture, and nutrient availability. Different nematode species have specific habitat preferences and ecological tolerances.

2.6.10 Ecological Functions

Nematodes play important roles in soil food webs and ecosystem functioning. They are key regulators of microbial populations and contribute to the decomposition of organic matter, nutrient cycling, and soil structure formation.

Overall, nematodes are integral components of soil ecosystems, contributing to soil health, nutrient cycling, and ecosystem resilience. Understanding the ecology and diversity of nematodes in soil are important for sustainable agriculture, ecosystem management, and conservation efforts.

2.7 Soil Arthropods

Soil arthropods are a diverse group of invertebrate organisms that inhabit soil ecosystems. They play vital roles in soil ecology, contributing to nutrient cycling, decomposition, soil structure formation, and regulation of microbial populations (Dervash et al. 2018). An overview of soil arthropods is as follows:

2.7.1 Diversity

Soil arthropods encompass a wide range of taxa, including insects, spiders, mites, centipedes, millipedes, and various other arthropods. They vary in size, shape, feeding habits, and ecological functions.

2.7.2 Feeding Strategies

On the basis of feeding habits, soil arthropods can be classified into the following:

2.7.2.1 Detritivores

These types of arthropods feed on decaying organic matter, such as plant litter, dead animals, and fungal hyphae. They facilitate decomposition and nutrient cycling by breaking down organic material into smaller particles and assisting microbial decomposition.

2.7.2.2 Predators

Predatory arthropods feed on other soil organisms, including bacteria, fungi, other arthropods, nematodes, and small vertebrates. They help control populations of other soil organisms and contribute to the regulation of soil food webs.

2.7.2.3 Herbivores

Some soil arthropods feed on living plant tissue, including roots, leaves, and stems. Herbivorous arthropods can have both beneficial and detrimental effects on plants, depending on their feeding habits and population densities.

2.7.2.4 Omnivores

Omnivorous arthropods have a mixed diet, consuming a variety of prey including detritus, fungi, bacteria, and other arthropods.

2.7.3 Soil Structure Formation

Soil arthropods, particularly burrowing species such as ants, termites, and earthworms, play important roles in soil structure formation. Their burrowing activities can help loosen and aerate the soil, improve water infiltration and drainage, and create microhabitats for other soil organisms.

2.7.4 Nutrient Cycling

Soil arthropods contribute to nutrient cycling and mineralization in soil ecosystems. By consuming organic matter and microbial biomass, they release nitrogen, phosphorus, and carbon in forms that are readily available to plants and other organisms (Dervash et al. 2018).

2.7.5 Interactions with Plants and Microorganisms

Soil arthropods interact with plants and soil microorganisms in various ways. For example, mycorrhizal fungi and bacteria may form relationships with the roots of certain plants, providing them with nutrients or protection in exchange for resources.

2.7.6   Environmental Factors

The abundance and activity of soil arthropods are affected by temperature, pH, soil moisture, and organic matter content. Different arthropod species have specific habitat preferences and ecological tolerances.

Overall, soil arthropods are integral components of soil ecosystems, contributing to soil health, nutrient cycling, and ecosystem resilience. Understanding the ecology and diversity of soil arthropods is important for sustainable agriculture, ecosystem management, and conservation efforts.

2.8   Soil Mesofauna

Microarthropods, also known as soil mesofauna, are a diverse group of invertebrates found within terrestrial samples, with sizes typically ranging from 0.1 to 2 mm (Swift et al. 1979). This taxonomic group includes various orders such as Acari (mites), collembolans (springtails), proturans, diplurans, symphellids, and enchytraeids (Dervash et al. 2018).

2.8.1   Diversity

They are dominant in most soils types, with a square meter of forest floor potentially containing numerous individuals signifying thousands of different species.

2.8.2   Habitat

Mites (Acari) are cosmopolitan in nature. Likewise, springtails (Collembola), are incredibly widespread and abundant in terrestrial ecosystems, occupying diverse habitats from polar regions to tropical forests (Potapov et al. 2024).

2.8.3   Environmental Factors

Their distribution within the soil profile can vary depending on temperature, organic matter content, soil moisture, and the presence of food resources.

2.8.4   Functions

Mesofauna are essential reservoirs of biodiversity, contributing to the overall richness and complexity of soil ecosystems. However, estimating species richness in
microarthropods, as with other soil organisms, presents challenges due to their
small size and cryptic nature. Soil mesofauna contribute to the functioning and stability of soil ecosystems through their diverse ecological roles. Their activities drive
nutrient cycling, decomposition, and soil structure maintenance, ultimately supporting plant growth and ecosystem efficiency (Dervash et al. 2018). Thus, they are well
known as ecosystem webmasters.

2.9   Earthworms in Soil

Earthworms are perhaps the most well-known and widely recognized soil organisms (Brown et al. 2023). They belong to the class Oligochaeta within the phylum
Annelida. Earthworms are highly beneficial to soil ecosystems and play dynamic
roles in soil health and fertility.

2.9.1   Diversity

There are numerous species of earthworms found in soil ecosystems worldwide,
although only a few species are commonly encountered in agricultural and garden
soils. These species vary in size, color, and ecological preferences (Ferreira
et al. 2023).

2.9.2   Habitat

Earthworms occupy a wide range of terrestrial habitats, for example, grasslands,
forests, agricultural fields, and gardens (Ferreira et al. 2023). They prefer soils with
high organic matter content, adequate moisture, and good drainage (Brown
et al. 2023).

There are several types of earthworms found in soil ecosystems, each with its
own characteristics and ecological roles (Kumar et al. 2023). Earthworms are generally categorized into three main ecological groups based on their behavior and habitat preferences:

(i) *Epigeic Earthworms.*

   Characteristics: Epigeic earthworms live in the top layers of soil, particularly in organic-rich environments such as leaf litter, compost heaps, and sur-

face litter layers. They are typically small to medium size and have relatively thin bodies (Kaur 2020).

Ecological Role: Epigeic earthworms are important decomposers, feeding on organic matter such as dead plant material and leaf litter. They play key roles in breaking down organic residues, accelerating decomposition, and recycling nutrients in surface soils (Kim et al. 2022).

Examples: It includes red worms (*Eisenia fetida*), Indian blue worm (*Perionyx excavatus*), and various species of composting earthworms.

(ii) *Endogeic Earthworms.*

Characteristics: Endogeic earthworms live in the mineral soil horizons, burrowing through the soil and creating horizontal burrows. They tend to have pale bodies and are adapted for life in the soil, with relatively small heads and reduced external pigmentation (Kim et al. 2022).

Ecological Role: Endogeic earthworms feed primarily on soil organic matter, ingesting soil particles and organic residues as they burrow. They contribute to soil structure improvement, soil aeration, and nutrient cycling in subsoil layers (Kaur 2020).

Examples: Common species include *Aporrectodea* spp., *Allolobophora* spp., and *Octolasion* spp.

(iii) *Anecic Earthworms.*

Characteristics: Anecic earthworms live in deep vertical burrows, but they also move to the soil surface to feed on surface organic matter (Kim et al. 2022). They have distinct pigmentation patterns, with darker dorsal surfaces and lighter ventral surfaces.

Ecological Role: Anecic earthworms play important roles in mixing organic and mineral soil layers, incorporating surface organic matter into deeper soil horizons, and enhancing nutrient availability for plants (Ahmed and Al-Mutairi 2022). Their deep burrows also improve soil drainage and water infiltration.

Examples: *Lumbricus terrestris* (commonly known as the common earthworm or night crawler) is a well-known example of an anecic earthworm.

These ecological categories are not rigid, and some earthworm species may exhibit characteristics of more than one group. Additionally, earthworm diversity can vary widely depending on geographical location, soil type, and environmental conditions (Vršič et al. 2021). Overall, earthworms are integral components of soil ecosystems, contributing to soil health, fertility, and ecosystem functioning through their diverse ecological roles.

2.10 Small Soil Mammals

Small soil mammals are a group of terrestrial mammals that spend a significant portion of their lives underground, either burrowing or living in soil habitats. These mammals play important functions in nutrient cycling, soil ecosystem dynamics, and soil structure formation (Augustine et al. 2023). Some examples of small soil mammals are as follows:

2.10.1 Moles

Moles are small, insectivorous mammals with cylindrical bodies, short limbs, and powerful forelimbs adapted for digging. They create elaborate tunnel systems underground, feeding primarily on earthworms, insect larvae, and other invertebrates found in soil. Moles are important ecosystem engineers, aerating the soil, mixing organic and mineral layers, and enhancing soil fertility through their burrowing activities formation (Augustine et al. 2023). However, their burrowing can also cause damage to lawns, gardens, and agricultural fields.

2.10.2 Voles

Voles, also known as field mice or meadow mice, are small rodents that inhabit grasslands, meadows, and agricultural fields (Giraudoux et al. 2020). They construct shallow burrows in the soil, often hidden beneath vegetation or debris. Voles are herbivorous and feed on grasses, herbs, roots, and seeds. They play roles in seed dispersal, soil disturbance, and nutrient cycling through their feeding and burrowing activities.

2.10.3 Shrews

Shrews are small, insectivorous mammals with long, pointed snouts and small eyes. They are active hunters, feeding on insects, earthworms, and other invertebrates found in soil and leaf litter. Shrews may create shallow tunnels or burrows in the soil, using them for shelter and protection from predators (Babyesiza et al. 2023). Their foraging activities contribute to soil aeration and nutrient cycling by redistributing organic matter and soil particles.

2.10.4 Gophers

Gophers are burrowing rodents that inhabit grasslands, pastures, and agricultural fields. They construct extensive tunnel systems underground, consisting of burrows for nesting, food storage, and foraging. Gophers are herbivorous and feed on roots, tubers, and other underground plant parts. Their burrowing activities can aerate the soil, mix organic and mineral layers, and enhance soil fertility. However, they are considered agricultural pests in some regions due to their nurturing habits and damage to crops (Babyesiza et al. 2023).

2.10.5   Pocket Gophers

Pocket gophers are a type of gopher characterized by fur-lined cheek pouches, which are meant to carry food and nesting material. They create characteristic mounds/piles of soil on the surface as they excavate their burrows underground (Britannica 2024). Pocket gophers are herbivorous and feed on a variety of plant roots and underground stems. Their burrowing activities can have noteworthy influences on soil structure, vegetation dynamics, and ecosystem processes. These small soil mammals perform vital roles in soil ecosystem functioning, including nutrient cycling, soil aeration, and habitat creation (Augustine et al. 2023). However, they can also have ecological and economic impacts, particularly when their populations become overabundant or when they come into conflict with human activities such as agriculture and landscaping.

2.11   Conservation of Soil Biota

Conserving soil biota is crucial in the current context of environmental challenges and agricultural demands. Soil biota underpins essential ecosystem services such as nutrient cycling, soil structure maintenance, and plant health. However, intensive agriculture, pollution, habitat destruction, and climate change pose significant threats to soil biota. Implementing sustainable land management practices, such as organic farming, reduced use of agrochemicals, agroforestry, and habitat restoration, is imperative. These strategies not only protect and enhance soil biodiversity but also ensure long-term soil health, agricultural productivity, and ecosystem resilience, which are vital for addressing global food security and environmental sustainability.

References

Abawi GS, Widmer TL (2000) Impact of soil health management practices on soilborne pathogens, nematodes and root diseases of vegetable crops. Appl Soil Ecol 15:37–47. https://doi.org/10.1016/S0929-1393(00)00070-6

Ahmed N, Al-Mutairi KA (2022) Earthworms effect on microbial population and soil fertility as well as their interaction with agriculture practices. Sustain For 14:7803. https://doi.org/10.3390/su14137803

Anabtawi HM, Lee WH, Al-Anazi A, Mohamed MM, Aly Hassan A (2024) Advancements in biological strategies for controlling harmful algal blooms (HABs). Water 16:224. https://doi.org/10.3390/w16020224

Augustine DJ, Smith JE, Davidson AD, Stapp P (2023) Burrowing rodents. In: McNew LB, Dahlgren DK, Beck JL (eds) Rangeland wildlife ecology and conservation. Springer, Cham. https://doi.org/10.1007/978-3-031-34037-6_15

Babyesiza WS, Mpagi J, Ssuuna J, Akoth S, Ectoparasite KA (2023) Fauna of rodents and shrews with their spatial, temporal, and dispersal along a degradation gradient in Mabira central Forest reserve. J Parasitol Res 2023:7074041

Bahram M, Netherway T (2022) Fungi as mediators linking organisms and ecosystems. FEMS Microbiol Rev 46:fuab058. https://doi.org/10.1093/femsre/fuab058

Battistuzzi M, Cocola L, Claudi R, Pozzer AC, Segalla A, Simionato D, Morosinotto T, Poletto L, La Rocca N (2023) Oxygenic photosynthetic responses of cyanobacteria exposed under an M-dwarf starlight simulator: implications for exoplanet's habitability. Front Plant Sci 14:1070359

Britannica T (2024) Editors of Encyclopaedia (2024, April 19). pocket gopher. Encyclopedia Britannica. https://www.britannica.com/animal/pocket-gopher

Brown GG, James SW, Csuzdi C, Lapied E, Decaëns T, Reynolds JW, Mısırlıoğlu M, Stovanic M, Trakić T, Sekulić J, Phillips H, Cameron E. (2023) A checklist of megadrile earthworm (Annelida: Clitellata) species and subspecies of the world. Available from: Zenodo. https://doi.org/10.5281/zenodo.7301848

Chamkhi I, El-Omari N, Balahbib A, El-Menyiy N, Benali T, Ghoulam C (2022) Is the rhizosphere a source of applicable multi-beneficial microorganisms for plant enhancement? Saudi J Biol Sci 29:1246–1259. https://doi.org/10.1016/j.sjbs.2021.09.032

Crouzet O, Consentino L, Pétraud JP, Marrauld C, Aguer JP, Bureau S, Le Bourvellec C, Touloumet L, Bérard A (2019) Soil photosynthetic microbial communities mediate aggregate stability: influence of cropping systems and herbicide use in an agricultural soil. Front Microbiol 10:1319. https://doi.org/10.3389/fmicb.2019.01319

Datta R (2024) Enzymatic degradation of cellulose in soil: a review. Heliyon 10:e24022. https://doi.org/10.1016/j.heliyon.2024.e24022

Dervash MA, Bhat RA, Mushtaq N, Singh DV (2018) Dynamics and importance of soil mesofauna. Int J Adv Res Sci Eng 7:2010–2019

Ebbisa A (2023) Arbuscular Mycorrhizal fungi (AMF) in optimizing nutrient bioavailability and reducing agrochemicals for maintaining sustainable agroecosystems: arbuscular Mycorrhizal fungi in agriculture—new insights. IntechOpen. https://doi.org/10.5772/intechopen.106995

Ferreira T, James SW, Bartz MLC, Lima ACR, Dudas R, Brown GG (2023) Distribution and diversity of earthworms in different land use systems in Rio Grande do Sul, Brazil. Zootaxa 5225:389–398. https://doi.org/10.11646/zootaxa.5225.1.31

Geisen S, Mitchell AED, Adl S, Bonkowski M, Dunthorn M, Ekelund F, Fernández LD, Jousset A, Krashevska V, Singer D, Spiegel FW, Walochnik J, Lara E (2018) Soil protists: a fertile frontier in soil biology research. FEMS Microbiol Rev 42:293–323. https://doi.org/10.1093/femsre/fuy006

Giraudoux P, Levret A, Afonso E, Coeurdassier M, Couval G (2020) Numerical response of predators to large variations of grassland vole abundance and long-term community changes. Ecol Evol 10:14221–14246. https://doi.org/10.1002/ece3.7020

Gonsior M, Hertkorn N, Hinman N (2018) Yellowstone hot springs are organic chemodiversity hot spots. Sci Rep 8:14155. https://doi.org/10.1038/s41598-018-32593-x

Gowtham HG, Hema P, Murali M, Shilpa N, Nataraj K, Basavaraj GL, Singh SB, Aiyaz M, Udayashankar AC, Amruthesh KN (2024) Fungal endophytes as mitigators against biotic and abiotic stresses in crop plants. J Fungi 10:116. https://doi.org/10.3390/jof10020116

Grzyb A, Wolna-Maruwka A, Niewiadomska A (2021) The significance of microbial transformation of nitrogen compounds in the light of integrated crop management. Agronomy 7:1415. https://doi.org/10.3390/agronomy11071415

Jassey VEJ, Hamard S, Lepère C, Céréghino R, Corbara B, Küttim M, Leflaiv J, Leroy C, Carrias JF (2022) Photosynthetic microorganisms effectively contribute to bryophyte CO_2 fixation in boreal and tropical regions. ISME Commun 2:64. https://doi.org/10.1038/s43705-022-00149-w

Kaur T (2020) Vermicomposting: an effective option for recycling organic wastes. IntechOpen. https://doi.org/10.5772/intechopen.91892

Khanum TA, Mehmood N, Khatoon N (2022) Nematodes as biological indicators of soil quality in the agroecosystems. IntechOpen. https://doi.org/10.5772/intechopen.99745

Kim G, Jo H, Kim HS, Kwon M, Son Y (2022) Earthworm effects on soil biogeochemistry in temperate forests focusing on stable isotope tracing: a review. Appl Biol Chem 65:88. https://doi.org/10.1186/s13765-022-00758-y

Kumar R, Yadav R, Gupta RK, Yodha K, Kataria SK, Kadyan P, Sharma P, Kaur S (2023) The earthworms: Charles Darwin's ecosystem engineer. IntechOpen. https://doi.org/10.5772/intechopen.1001339

Lamb JA, Fernandez FG, Kaiser DE (2014) Understanding nitrogen in soils. University of Minnesota, Minneapolis, pp 1–5

Li X, Liu C, Zhao H, Gao F, Ji G, Hu F, Li H (2018) Similar positive effects of beneficial bacteria, nematodes and earthworms on soil quality and productivity. Appl Soil Ecol 130:202–208

Lindström K, Mousavi SA (2020) Effectiveness of nitrogen fixation in rhizobia. Microb Biotechnol 13:1314–1335. https://doi.org/10.1111/1751-7915.13517

Lladó S, López-Mondéjar R, Baldrian P (2017) Forest soil bacteria: diversity, involvement in ecosystem processes, and response to global change. Microbiol Mol Biol 81:e00063–e00016. https://doi.org/10.1128/mmbr.00063-16

Molefe RR, Amoo AE, Babalola OO (2023) Communication between plant roots and the soil microbiome; involvement in plant growth and development. Symbiosis 90:231–239. https://doi.org/10.1007/s13199-023-00941-9

Neher DA, Barbercheck ME (2019) Soil microarthropods and soil health: intersection of decomposition and pest suppression in agroecosystems. Insects 10:414. https://doi.org/10.3390/insects10120414

Ozturk M, Khan S, Egamberdieva D, Khassanov F, Efe R, Altay V (2022) Biodiversity, conservation and sustainability in Asia—volume 2: prospects and challenges in south and middle Asia. Springer Nature, p 1070

Policelli N, Horton TR, Hudon AT, Patterson TR, Bhatnagar JM (2020) Back to roots: the role of ectomycorrhizal fungi in boreal and temperate forest restoration. Front For Glob Change 3:97. https://doi.org/10.3389/ffgc.2020.00097

Potapov AM, Chen TW, Striuchkova AV et al (2024) Global fine-resolution data on springtail abundance and community structure. Sci Data 11:22. https://doi.org/10.1038/s41597-023-02784-x

Robb C (2024) The mainstreaming agenda of the convention on biological diversity and its value to protecting and enhancing soil ecosystem services. In: Ginzky H et al (eds) International yearbook of soil law and policy. Springer, Cham. https://doi.org/10.1007/978-3-031-40609-6_8

Solomon W, Mutum L, Janda T, Molnár Z (2023) Potential benefit of microalgae and their interaction with bacteria to sustainable crop production. Plant Growth Regul 101:53–65. https://doi.org/10.1007/s10725-023-01019-8

Swift MJ, Heal OW, Anderson JM (1979) Decomposition in terrestrial ecosystems. University of California Press, Berkeley

Tian J, Ge F, Zhang D, Deng S, Liu X (2021) Roles of phosphate solubilizing microorganisms from managing soil phosphorus deficiency to mediating biogeochemical P cycle. Biology (Basel) 10:158. https://doi.org/10.3390/biology10020158

Tibbett M, Fraser TD, Duddigan S (2020) Identifying potential threats to soil biodiversity. PeerJ 12:e9271. https://doi.org/10.7717/peerj.9271

Vršič S, Breznik M, Pulko B, Rodrigo-Comino J (2021) Earthworm abundance changes depending on soil management practices in Slovenian vineyards. Agronomy 11:1241. https://doi.org/10.3390/agronomy11061241

Wall DH, Nielsen UN (2012) Biodiversity and ecosystem services: is it the same below ground? Nature Edu Knowledge 3:8

Yao X, Guo H, Zhang K, Zhao M, Ruan J, Chen J (2023) Trichoderma and its role in biological control of plant fungal and nematode disease. Front Microbiol 14:1160551. https://doi.org/10.3389/fmicb.2023.1160551

Zeng X, Gao H, Wang R, Majcher BM, Woon JS, Wenda C, Eggleton P, Griffiths HM, Ashton LA (2024) Global contribution of invertebrates to forest litter decomposition. Ecol Lett 27:4

Chapter 3
Soil Stress Ecology: *Concept, Impacts, and Management Strategies*

Abstract Stress ecology of soil examines how various stressors (wide range of abiotic and biotic stressors) impact soil ecosystems, including their physical, chemical, and biological properties, and subsequently affect plant and microbial communities. Soil interacts dynamically with environmental factors, and these interactions influence its functionality and ecological balance. Understanding soil stress ecology is crucial for enhancing soil health, agricultural productivity, and environmental sustainability. The stress ecology of soil urges on maintaining soil ecosystem balance amid natural and anthropogenic pressures. Stressors can negatively affect soil biota, altering biodiversity, ecosystem functioning, and essential agricultural services such as nutrient cycling, soil fertility, and pest regulation, as detailed in this chapter.

3.1 Introduction

Stress ecology of soil focuses on understanding how various stressors affect soil ecosystems, including their physical, chemical, and biological properties, and how these modifications, in turn, affect plant and microbial communities within the soil. Soil, as a living system, interacts dynamically with environmental factors and the stresses they impose, impacting its function and the broader ecological balance. Understanding and addressing the stress ecology of soil not only aid in enhancing soil health and agricultural productivity but also support broader environmental sustainability goals. The focus is on maintaining the delicate balance of soil ecosystems in the face of both natural and anthropogenic pressures. Soil stressors can have detrimental effects on soil biota, altering soil biodiversity, ecosystem functioning, and the provision of ecosystem services essential for agriculture, such as nutrient cycling, soil fertility, and pest regulation which is comprehensively addressed in this chapter.

3.2   Abiotic Stressors and Soil Microbiome

Soil microbes play crucial roles in nutrient cycling and organic matter decomposition. Abiotic stress can disrupt these processes by altering microbial metabolism. For example, high temperatures can accelerate microbial activity, leading to increased decomposition rates, whereas extreme pH levels can inhibit certain enzymatic activities. Abiotic stressors (extreme temperatures, salinity, and drought) can significantly impact the soil microbiome, leading to various effects on microbial communities and ecosystem functioning (Fadiji et al. 2023). But, contrarily, abiotic stresses can indeed trigger plants to produce bioactive compounds as a defense mechanism. These bioactive compounds often have beneficial effects on human health and are known as nutraceuticals. By strategically leveraging these stresses, we can potentially enhance the nutritional quality and health-promoting properties of crops. However, it's essential to balance stress levels carefully to avoid compromising crop yield and overall plant health. Proper management and understanding of plant responses to stress are crucial for maximizing the production of bioactive compounds while maintaining crop productivity.

The specific examples include the following:

3.2.1   Chemical Stress on Soil

Chemical stressors include pollutants, salinity, and changes in pH:

(a) *Environmental Pollutants*: Heavy metals, pesticides, and other contaminants can inhibit microbial activity and affect plant health. Heavy metals can accumulate in soils due to industrial activities, mining, or improper waste disposal. Metal stress can alter microbial community structure and function, affecting important soil processes like carbon and nitrogen cycling (Angon et al. 2024). Metal-contaminated soils can harbor metal-resistant microbial communities capable of detoxifying or sequestering heavy metals through processes like bioremediation (Ghori et al. 2019).

(b) *Salinity*: High soil salinity levels can select for halophilic (salt-tolerant) microbial species while suppressing the growth of sensitive organisms. Salinity can disrupt microbial-mediated processes like nitrogen fixation and phosphorus solubilization, leading to nutrient imbalances in the soil (Meinzer et al. 2023; Ozturk et al. 2006). Certain bacteria and archaea, such as *Halomonas* and *Haloarchaea*, thrive in saline environments and contribute to salt tolerance in soils.

(c) *pH Changes (Acidification and Alkalization)*: Changes in soil pH, either due to natural processes or anthropogenic activities like acid rain deposition, can impact microbial community composition and metabolic activities (Wang et al. 2022). Acidic soils may favor acidophilic microorganisms like *Acidobacteria,* while alkaline soils may select for alkali-tolerant species such as certain

Actinobacteria. The pH extremes can influence the availability of nutrients and the solubility of toxic elements, indirectly affecting microbial communities.

3.2.2 Hydrological Stress on Soil

(a) *Drought*: Drought conditions reduce soil moisture, which can lead to a decline in the abundance and diversity of soil microbial communities. Reduced water availability can impair microbial activity involved in nutrient cycling, such as nitrogen fixation and organic matter decomposition. Drought-tolerant species such as *Actinobacteria* and certain fungal taxa may become more dominant, while moisture-sensitive species decline (Bogati and Walczak 2022).

(b) *Waterlogging*: Waterlogged or anaerobic conditions in soils restrict oxygen availability, altering microbial community composition and functioning (Ansari et al. 2023). Anaerobic microbes such as sulfate-reducing bacteria and methanogenic archaea may become more abundant in waterlogged soils (Kotsyurbenko et al. 2019). Waterlogging can lead to the accumulation of toxic metabolites like hydrogen sulfide and methane, affecting soil health and greenhouse gas emissions (Ashraf et al. 2009; Ozturk et al. 2006).

3.2.3 Physical Stress on Soil

Physical stress on soil includes compaction and erosion:

(a) *Compaction*: Reduce soil porosity, limiting water infiltration and root growth. Changes in the soil microbiome due to compaction can have feedback effects on soil health and productivity. For example, alterations in microbial community structure can influence soil structure, water retention capacity, and susceptibility to erosion (Frene et al. 2024).

(b) *Soil erosion*: Removal of topsoil by wind or water reduces soil fertility and structure. Erosion can disrupt nutrient cycling processes in soil by affecting microbial communities involved in nitrogen fixation, phosphorus solubilization, and other important functions. This can lead to nutrient imbalances and reduced soil fertility (Qiu et al. 2021).

3.2.4 Meteorological Stress

Meteorological stress in soil refers to the impact of weather and climate elements such as temperature, precipitation, wind, and atmospheric composition on soil properties and processes. These stresses can significantly influence soil structure, chemistry, biology, and overall ecosystem functions (Gelybó et al. 2018):

3.2.4.1   Temperature Fluctuations

Temperature extremes and fluctuations can have profound effects on soil health as follows:

(i) *High Temperatures*: Can lead to increased soil evaporation rates, reduced soil moisture, and altered microbial activity. It can also upsurge the rate of organic matter decomposition, reducing soil fertility over time.
(ii) *Low Temperatures*: Freeze-thaw cycles can affect soil structure by promoting soil aggregation and, conversely, soil dispersion. This can impact water infiltration and root penetration.

3.2.4.2   Precipitation Variability

Changes in precipitation patterns can induce various forms of stress on soils. Reduced rainfall leads to lower soil moisture, impacting plant growth and microbial activities crucial for nutrient cycling (Li et al. 2022). Contrarily, heavy precipitation leads to soil erosion, leaching of nutrients, and, in severe cases, flooding. This can strip away the fertile top layer of soil and disturb the balance of the soil ecosystem.

3.2.4.3   Wind

Strong winds can contribute to soil erosion, particularly in arid and semiarid regions. Wind erosion removes topsoil containing organic matter and nutrients, leading to degradation of soil quality and reduced agricultural productivity.

3.2.4.4   Atmospheric Deposition

Modifications in the chemical composition of the atmosphere can affect soil chemistry as follows:

(i) *Acid Rain*: Sulfur dioxide and nitrogen oxides from industrial activities can dissolve into rainwater to form acids. Acid deposition can lower soil pH, which in turn can mobilize toxic metals and affect nutrient availability (Prakash et al. 2023).
(ii) *Pollutants*: Other airborne pollutants, including particulates from industrial processes or volcanic ash, can also deposit on soils affecting their chemical properties and potentially leading to contamination.

3.2.4.5   Extreme Weather Events

Climate change escalates the frequency and intensity of extreme weather events, which can have severe impacts on soil as follows:

 (i) *Storms and Hurricanes*: Can lead to significant erosion and waterlogging.
(ii) *Heat waves*: Prolonged high temperatures can lead to soil desiccation, making
     it less productive and reducing its ability to support plant life.

3.2.4.6 Snow and Ice Cover

Extended periods of snow and ice can impact soil processes as follows:

 (i) *Insulation*: Snow can insulate the soil, affecting the freezing depth and the
     activities of soil organisms.
(ii) *Melting Patterns*: Uneven melting can contribute to soil erosion and nutri-
     ent runoff.

3.2.4.7 Management-Induced Stresses

Agricultural and land management practices often induce stress as follows:

(a) *Overgrazing*: Can lead to soil degradation and increased erosion.
(b) *Overcultivation*: Reduces soil fertility through the loss of organic matter.
(c) *Pesticide and Fertilizer Use*: Can lead to chemical imbalances and pollution.

3.3 Biological Stress and Soil Microbiome

Biotic soil stress refers to challenges faced by soil organisms due to interactions
with other life-forms, such as plants, animals, and other microbes. These interactions can impact the composition, activity of soil microbiomes and diversity, as well
as ecosystem functioning (Ho et al. 2017). Some examples of biotic soil stressors
include the following:

3.3.1 Microbial Interactions

Soil microbes engage in complex interactions with each other, including competition, predation, mutualism, and parasitism, which can influence their abundance,
diversity, and functions (Wu et al. 2023). Competition for nutrients, space, and other
resources among soil microbes can lead to shifts in community structure and dynamics, especially under resource-limiting conditions. Mutualistic interactions between
plants and soil microbes, such as mycorrhizal associations and nitrogen-fixing sym-
bioses, can alleviate biotic stress and enhance plant productivity.

3.3.2 Plant Exudates and Allelopathy

Plants release a range of organic compounds into the soil, collectively known as root exudates. These exudates can act as both stressors and resources for soil microbes. The composition and magnitude of root exudates can differ depending on plant species, developmental stage, and environmental conditions (Sharma et al. 2023). Some plants produce allelochemicals that inhibit the germination, growth, or reproduction of other plants and soil microbes, a phenomenon known as allelopathy. Allelochemicals released from plant residues or root exudates can impact the structure and functioning of soil microbial communities, potentially affecting nutrient cycling and ecosystem processes.

3.3.3 Pathogens and Pests

Diseases and pests can reduce plant health and affect nutrient cycling. Soilborne pathogens, such as certain fungi, bacteria, nematodes, and viruses, can infect plant roots and cause diseases, leading to stress for both plants and associated soil microbes (Panth et al. 2020). Plant pathogens can alter root exudation patterns, suppress plant immune responses, and modify soil microenvironments, impacting the abundance and diversity of soil microbiomes. Some soil microbes may act as antagonists or biocontrol agents against phytopathogens, helping to mitigate their negative effects on plant health (Kundoo et al. 2021).

3.3.4 Root Herbivory

Soil-dwelling herbivorous organisms, such as root-feeding insects, nematodes, and rodents, can damage plant roots and alter root exudation profiles. Root herbivory can induce plant defense responses, including changes in root exudate composition and release of signaling molecules that affect soil microbial communities. Soil microbes may respond to root herbivory by modulating their metabolic activities, competing for resources, or forming symbiotic associations with plants to enhance resistance against herbivores (Schandry and Becker 2020).

3.3.5 Invasive Species

These can change soil structure, nutrient cycling, and the competitive landscape for native species.

These examples highlight the diverse ways in which biotic interactions can shape soil microbial communities and ecosystem dynamics, emphasizing the interconnectedness of aboveground and belowground processes in terrestrial ecosystems. Understanding the role of biotic soil stressors is essential for elucidating the mechanisms driving soil biodiversity, ecosystem functioning, and resilience to environmental changes.

3.4 Vascular Plants, Microbes, and Greenhouse Gas Emissions

Vascular plants, equipped with lignified tissues, are pivotal in terrestrial environments as they facilitate the movement of water, minerals, and photosynthetic substances. They stand as the primary producers in ecosystems, absorbing substantial amounts of atmospheric carbon dioxide (CO_2) during photosynthesis. Yet, not all the absorbed CO_2 remains stored; some is respired by the plant itself, while additional CO_2 is emitted from rhizodeposits. These deposits are broken down by soil microorganisms associated with the plant's roots (Gul and Whalen 2013).

Microbial processes play a significant role in the conversion of organic nitrogen (N) derived from plant remnants (such as rhizodeposits and dead plant material) into nitrous oxide (N_2O), a potent greenhouse gas, through mineralization and subsequent nitrification. Additionally, in anaerobic soil conditions, nitrate reduction by denitrifying microorganisms also contributes to N_2O production. The intricate interactions between plants and microbes in soil, which lead to emissions of CO_2 and N_2O, could be influenced by genetic modifications. For instance, reducing the expression of genes involved in lignin biosynthesis to lower lignin levels or alter the guaiacyl-syringyl (G:S) ratio in aboveground biomass holds potential benefits such as enhanced digestibility in forage crops, improved suitability for the wood pulping industry in short rotation woody crops, and the development of second-generation biofuel crops with reduced lignocellulosic content. However, the rapid decomposition of unharvested residues from such genetically modified crops may lead to increased CO_2 and N_2O emissions from the soil (Gul and Whalen 2013).

3.4.1 Case Study of Genetically Modified Cell Wall Mutants: Arabidopsis Thaliana

Gul and Whalen (2013) conducted a study to investigate the impact of altering plant lignin biosynthesis on soil CO_2 and N_2O emissions, utilizing experimental findings from genetically modified cell wall mutants of *Arabidopsis thaliana*. The study employs conceptual models illustrating how modifications in lignin biosynthesis affect the timing of plant growth stages, physical characteristics, and biomass output, consequently influencing the distribution of photosynthetic resources and carbon losses via rhizodeposition and respiration throughout the plant's life cycle, as well as the chemical makeup of plant residues.

The interconnected impacts of soil conditions (including mineral nitrogen levels, soil moisture, microbial communities, and aggregation) on the release of CO_2 and N_2O are outlined. While targeting the downregulation of the cinnamoyl-CoA reductase 1 (CCR1) gene holds promise for developing highly digestible forage and biofuel crops, *Arabidopsis thaliana* with this mutation exhibits reduced plant biomass and fertility, prolonged vegetative growth, and plant residues that degrade more readily, resulting in increased short-term emissions of CO_2 and N_2O from the soil. The future challenge in crop breeding endeavors will be to identify tissue-specific genes related to lignin production that fulfill market demands without compromising objectives related to soil CO_2 and N_2O emissions, as highlighted by Gul and Whalen (2013).

3.5   Ecological and Functional Impacts

Stress on soil ecosystems can have profound effects on their ecological and functional capacities as follows:

3.5.1   Biodiversity Loss

Soil stress can reduce both plant and microbial diversity, which is crucial for ecosystem resilience.

3.5.2   Alteration of Nutrient Cycles

Stress can disrupt the cycles of carbon, nitrogen, and other essential elements.

3.5.3   Reduced Productivity

Stressed soils often show diminished productivity, impacting agriculture and natural vegetation.

3.6   Soil Stress and Physiological Changes

In stressful conditions, such as drought or heat stress, plants often experience physiological changes as they adapt to the challenging environment. These changes can influence various aspects of fruit development, including sugar content and sweetness. The few reasons why fruits may become sweeter under stress conditions are as under:

3.6.1 Osmotic Adjustment

Under water stress, plants may implement osmotic adjustment mechanisms to maintain cell turgor pressure and water uptake (Yang et al. 2021). This adjustment often involves accumulating solutes, such as sugars, in the fruit cells. The higher concentration of sugars within the fruit can result in increased sweetness.

3.6.2 Reduced Water Content

Water stress can result in reduced water uptake by the plant, which may result in smaller fruit size (Lima et al. 2023). With less water content diluting the sugar concentration, the sweetness of the fruit can be more pronounced.

3.6.3 Limited Growth

Stress conditions may limit overall plant growth and development, including fruit growth (Lima et al. 2023). In some cases, when fruit growth is restricted, the sugars produced by photosynthesis are distributed into a smaller fruit, leading to higher sugar concentration and increased sweetness.

3.6.4 Altered Metabolism

Environmental stress can trigger changes in plant metabolism, directing resources toward pathways that produce osmolytes, including sugars (Jogawat et al. 2021). This metabolic shift can lead to the accumulation of sugars in fruits, enhancing their sweetness.

3.6.5 Hormonal Regulation

Stress conditions can also influence hormone levels in plants, such as abscisic acid (ABA), which plays a role in plant stress responses (Sah et al. 2016). ABA can stimulate the production and accumulation of sugars in fruits, contributing to their sweetness.

3.6.6 Survival Strategy

Sweet fruits may serve as a strategy for plants to attract seed dispersers, such as animals, even under stress conditions. By producing sweeter fruits, plants may increase the likelihood of seed dispersal and enhance their chances of survival and reproduction.

3.6.7 Genetic and Species Variability

Different plant species and cultivars may respond differently to stress conditions. Some may increase sugar production under stress, while others may not show significant changes. Genetic variability and adaptation play a role in determining the specific response of each plant to stress (Raza et al. 2024).

3.6.8 Increase in Carotenoids and Sugars

Carotenoids are pigments that contribute to fruit coloration and also act as antioxidants, helping to protect the plant from damage caused by reactive oxygen species (ROS) generated during stress. Sugars serve as osmolytes, helping to maintain cellular water balance and stabilize cell structures under stress conditions. When plants experience stress, they often increase the production of certain compounds like carotenoids and sugars and have been reported by Akrami and Arzani (2018) in melon. But, at the same time, malondialdehyde (MDA) and hydrogen peroxide (H_2O_2) are markers of oxidative stress, indicating damage to cell membranes and increased production of ROS, respectively (García-Caparrós et al. 2020). Elevated levels of these compounds suggest that the plant is experiencing oxidative damage due to salinity stress. Changes in chlorophyll content reflect alterations in photosynthetic activity, which can be disrupted by salinity stress. Proline accumulation is a common response to various environmental stresses and serves as a compatible solute, helping to maintain cell turgidity and protect proteins and enzymes from denaturation (Toscano et al. 2019).

While fruit sweetness can increase under stress conditions, it's essential to note that prolonged or severe stress can have detrimental effects on overall plant health and productivity. Therefore, while sweeter fruits may be observed under certain stress conditions, efforts to mitigate stress through proper irrigation, soil management, and other cultivation practices are crucial for sustaining long-term crop production and quality.

3.7 Impacts of Soil Stress on Soil Biota

Soil stress can have significant impacts on soil organisms, disrupting their populations, diversity, and activities, which in turn can affect various ecosystem processes.
Some of the key impacts of soil stress on soil organisms include the following:

3.7.1 Changes in Population Dynamics

Soil stressors such as pollution, compaction, or extreme temperatures can directly
affect the survival and reproduction of soil organisms. This can lead to shifts in
population sizes and changes in the relative abundance of different species within
soil communities.

3.7.2 Loss of Biodiversity

Soil stressors can disproportionately affect certain groups of soil organisms, leading
to declines in species diversity. Loss of biodiversity can weaken ecosystem resilience and reduce the ability of soil communities to perform essential functions such
as decomposition and nutrient cycling.

3.7.3 Altered Community Composition

Soil stress can favor the growth and proliferation of stress-tolerant species while
suppressing sensitive ones. This can result in changes in the structure and composition of soil communities, potentially leading to imbalances in ecosystem functioning.

3.7.4 Impaired Ecosystem Processes

Soil organisms are involved in crucial ecosystem processes such as soil formation,
decomposition, and nutrient cycling. Soil stress can disrupt these processes by
reducing the activity and efficiency of soil organisms, ultimately affecting the productivity and stability of ecosystems.

3.7.5   Loss of Ecosystem Services

Soil organisms provide various ecosystem services that are indispensable for human well-being, including carbon sequestration, soil fertility maintenance, and pest regulation. Soil stress can compromise the provision of these services, leading to negative impacts on agriculture, food safety assurance, and environmental quality.

3.7.6   Feedback Effects on Soil Health

Alterations in soil organism communities resulting from soil stress can further exacerbate soil degradation and stress. For example, declines in soil biodiversity can reduce soil resilience to future disturbances, creating a feedback loop of degradation and decline.

Therefore, understanding the impact of soil stress on soil organisms is important for predicting and mitigating the consequences of environmental change on soil ecosystems and the services they provide. Conservation and restoration efforts aimed at promoting soil health and resilience can help mitigate these impacts and maintain the functionality of soil ecosystems.

3.8   Management and Mitigation Strategies

To manage and mitigate the impact of stress on soils, several sustainable strategies can be adopted as follows:

(a) Techniques such as cover cropping, contour farming, and maintaining vegetation cover can help reduce erosion, maintain soil moisture, and reduce pest and disease cycles.

(b) Efficient irrigation practices and water retention strategies (like swales and rain gardens) can help manage both drought and excessive rainfall. Efficient irrigation systems like drip or sprinkler irrigation can help manage water use more effectively, reducing water stress and preventing salinization.

(c) Adding organic matter through compost or green manure can improve soil structure, enhance water retention, and increase resilience against temperature and moisture stresses.

(d) Conservation tillage reduces erosion and maintains soil structure.

(e) Combined biological, cultural, mechanical, and chemical management practices (integrated pest management) to control pests and diseases can minimize the reliance on chemical pesticides.

(f) Restoration of degraded soils through techniques like reforestation, biocomposting, and biochar application to enhance soil organic matter.

(g) Regular soil testing can guide the balanced application of fertilizers, optimizing nutrient availability and minimizing excesses or deficiencies.

By understanding and managing these stress factors, farmers can substantially improve the health of their soils and the sustainability of their agricultural practices, leading to better crop yields and the long-term viability of their farming operations.

References

Akrami M, Arzani A (2018) Physiological alterations due to field salinity stress in melon (Cucumis melo L.). Acta Physiol Plant 40:1–14. https://doi.org/10.1007/s11738-018-2657-0

Angon PB, Islam MS, Shreejana KC, Das A, Anjum N, Poudel A, Suchi SA (2024) Sources, effects and present perspectives of heavy metals contamination: soil, plants and human food chain. Heliyon 10:e28357. https://doi.org/10.1016/j.heliyon.2024.e28357

Ansari MKA, Unal BT, Javad S, Vardar F, Ansari AA, Ozturk M, Iqbal M (2023) Application of nanobiotechnology in enabling plants to overcome water-logging stress: a review. Agric Rev 43:289–299

Ashraf M, Ozturk M, Athar HR (eds) (2009) Salinity and water stress: improving crop efficiency. (Tasks for Vegetation Science 44). Springer Science+Business Media

Bogati K, Walczak M (2022) The impact of drought stress on soil microbial community, enzyme activities and plants. Agronomy 12:189. https://doi.org/10.3390/agronomy12010189

Fadiji AE, Yadav AN, Santoyo G, Babalola OO (2023) Understanding the plant-microbe interactions in environments exposed to abiotic stresses: an overview. Microbiol Res 271:127368. https://doi.org/10.1016/j.micres.2023.127368

Frene JP, Pandey BK, Castrillo G (2024) Under pressure: elucidating soil compaction and its effect on soil functions. Plant Soil. https://doi.org/10.1007/s11104-024-06573-2

García-Caparrós P, DeFilippis L, Gul A, Hasanuzzaman M, Ozturk M, Altay V, Lao MT (2020) Oxidative stress and antioxidant metabolism under adverse environmental conditions: a review. BotRev. https://doi.org/10.1007/s12229-020-09231-1

Gelybó G, Tóth E, Farkas C, Agota H, Kása I, Bakacsi Z (2018) Potential impacts of climate change on soil properties. Agrokém Talajt 67:121–141. https://doi.org/10.1556/0088.2018.67.1.9

Ghori NH, Ghori T, Hayat MQ, Imadi SR, Gul A, Altay V, Ozturk M (2019) Heavy metal stress and responses in plants. Int J Environ Sci Technol 16:1807–1828

Gul S, Whalen J (2013) Plant life history and residue chemistry influences emissions of CO_2 and N_2O from soil—perspectives for genetically modified Cell Wall mutants. Crit Rev Plant Sci 32(5):344–368. https://doi.org/10.1080/07352689.2013.781455

Ho YN, Mathew DC, Huang CC (2017) Plant-microbe ecology: interactions of plants and symbiotic microbial communities. INTECH. https://doi.org/10.5772/intechopen.69088

Jogawat A, Yadav B, Chhaya LN, Singh AK, Narayan OP (2021) Crosstalk between Phytohormones and secondary metabolites in the drought stress tolerance of crop plants: a review. Physiol Plant 172:1106–1132

Kotsyurbenko OR, Glagolev MV, Merkel AY, Sabrekov AF, Terentieva IE (2019) Methanogenesis in soils, wetlands, and peat. In: Stams A, Sousa D (eds) Biogenesis of hydrocarbons. Handbook of hydrocarbon and lipid microbiology. Springer, Cham. https://doi.org/10.1007/978-3-319-78108-2_9

Kundoo AA, Dervash MA, Bhat RA, Hussain B, Mushtaq N (2021) Biocontrol agents in organic agriculture. In: Bhat RA, Hakeem KR, Qadri H, Dervash MA (eds) Agricultural waste: threats and technologies for its sustainable management. CRC Press

Li X, Yan Y, Lu X, Fu L, Liu Y (2022) Responses of soil bacterial communities to precipitation change in the semiarid alpine grassland of Northern Tibet. Front Plant Sci 13:1036369. https://doi.org/10.3389/fpls.2022.1036369

Lima JS, Andrade OVS, Morais EG, Machado GGL, Santos LC, Andrade ES, Benevenute PAN, Martins GS, Nascimento VL, Marchiori PER et al (2023) KI increases tomato fruit quality and water deficit tolerance by improving antioxidant enzyme activity and amino acid accumulation: a priming effect or relief during stress? Plan Theory 12:4023. https://doi.org/10.3390/plants12234023

Meinzer M, Ahmad N, Nielsen BL (2023) Halophilic plant-associated bacteria with plant growth promoting potential. Microorganisms 11:2910. https://doi.org/10.3390/microorganisms11122910

Ozturk M, Waisel Y, Khan MA, Gork G (eds) (2006) Biosaline agriculture and salinity tolerance in plants. Birkhauser Verlag (Springer Science), Basel

Panth M, Hassler SC, Baysal-Gurel F (2020) Methods for management of soilborne diseases in crop production. Agriculture 10:16. https://doi.org/10.3390/agriculture10010016

Prakash J, Agrawal SB, Agrawal M (2023) Global trends of acidity in rainfall and its impact on plants and soil. J Soil Sci Plant Nutr 23(1):398–419. https://doi.org/10.1007/s42729-022-01051-z

Qiu L, Zhang Q, Zhu H, Reich PB, Banerjee S, van der Heijden MGA, Sadowsky MJ, Ishii S, Jia X, Shao M, Liu B, Jiao H, Li H, Wei X (2021) Erosion reduces soil microbial diversity, network complexity and multi-functionality. ISME J 15:2474–2489. https://doi.org/10.1038/s41396-021-00913-1

Raza A, Bhardwaj S, Rahman MD, García-Caparrós P, Habib M, Saeed F, Charagh S, Foyer HC, Siddique KHM, Varshney RK (2024) Trehalose: a sugar molecule involved in temperature stress management in plant. Crop J 12:1–16. https://doi.org/10.1016/j.cj.2023.09.010

Sah SK, Reddy KR, Li J (2016) Abscisic acid and abiotic stress tolerance in crop plants. Front Plant Sci 7:571

Schandry N, Becker C (2020) Allelopathic plants: models for studying plant–interkingdom interactions. Trends Plant Sci 25:176–185. https://doi.org/10.1016/j.tplants.2019.11.004

Sharma I, Kashyap S, Agarwala N (2023) Biotic stress-induced changes in root exudation confer plant stress tolerance by altering rhizospheric microbial community. Front Plant Sci 14:1132824. https://doi.org/10.3389/fpls.2023.1132824

Toscano S, Trivellini A, Cocetta G, Bulgari R, Francini A, Romano D, Ferrante A (2019) Effect of preharvest abiotic stresses on the accumulation of bioactive compounds in horticultural produce. Front Plant Sci 10:1212. https://doi.org/10.3389/fpls.2019.01212

Wang T, Cao X, Chen M, Lou Y, Wang H, Yang Q, Pan H, Zhuge Y (2022) Effects of soil acidification on bacterial and fungal communities in the Jiaodong peninsula, Northern China. Agronomy 12:927. https://doi.org/10.3390/agronomy12040927

Wu D, Wang W, Yao Y, Li H, Wang Q, Niu B (2023) Microbial interactions within beneficial consortia promote soil health. Sci Tot Environ 900:165801. https://doi.org/10.1016/j.scitotenv.2023.165801

Yang X, Lu M, Wang Y, Wang Y, Liu Z, Chen S (2021) Response mechanism of plants to drought stress. Horticulturae 7:50. https://doi.org/10.3390/horticulturae7030050

Chapter 4
Agricultural Industry: *Blessing and a Curse for Soil Biota*

Abstract The agricultural industry is a multifaceted sector involving crop cultivation, livestock raising, and the production of food, fiber, and other commodities. It plays a crucial role in the global economy by providing essential goods, raw materials, and livelihoods for millions worldwide. While it has significant positive impacts, it also poses challenges to society, the economy, and the environment. Balancing agricultural productivity with environmental sustainability and social equity requires integrated approaches that prioritize regenerative practices, conservation efforts, and inclusive policies. The odds and eves of agricultural enterprises on soil biota are discussed in this chapter.

4.1 Introduction

The agricultural industry is a complex and multifaceted sector which encompasses all accomplishments associated with the cultivation of crops, raising of livestock, and production of food, fiber, and other agricultural commodities. It is a vital sector of the global economy, providing essential goods for human consumption, raw materials for industries, and livelihoods for millions of people worldwide (Pawlak and Kołodziejczak 2020). However, it has both positive and negative impacts on society, economy, and environment. Achieving a balance between agricultural productivity, environmental sustainability, and social equity requires integrated approaches that prioritize regenerative practices, conservation efforts, and inclusive policies to ensure a resilient and equitable food system for future generations. Agricultural industry comes up with both positive and negative impacts on soil biodiversity, which are discussed in this chapter.

© The Author(s), under exclusive license to Springer Nature Switzerland AG 2024   39
M. A. Dervash et al., *Soil Organisms*, SpringerBriefs in Microbiology,
https://doi.org/10.1007/978-3-031-66293-5_4

4.2   Physiognomies of Agricultural Industry

The physiognomies of the agricultural industry encompass a wide range of aspects, from its structure and practices to its impact on society and the environment as follows:

4.2.1   Crop Production

Crop production involves the cultivation of various food crops, cash crops, and industrial crops on agricultural land. Major food crops include grains (e.g., rice, wheat, corn), oilseeds (e.g., soybeans, sunflowers), fruits, vegetables, and pulses (e.g., beans, lentils). Cash crops such as cotton, sugarcane, tobacco, and coffee are grown for commercial purposes, while industrial crops like rubber, jute, and sisal are used for manufacturing purposes.

4.2.2   Livestock Farming

Livestock farming involves the raising of domesticated animals for meat, milk, eggs, wool, leather, and other products. Common livestock species include cattle, poultry (chickens, ducks, turkeys), pigs, sheep, goats, and aquaculture species (fish, shrimp). Livestock farming practices vary widely, ranging from extensive grazing systems to intensive confinement operations.

4.2.3   Agribusiness

Agribusiness refers to the interconnected network of businesses and industries involved in agricultural production, processing, distribution, marketing, and retailing. It includes input suppliers (e.g., seed companies, agrochemical manufacturers), agricultural machinery and equipment manufacturers, food processors, wholesalers, retailers, and agri-food exporters/importers.

4.2.4   Supply Chain Management

The agriculture industry relies on complex supply chains to transport agricultural inputs (e.g., seeds, fertilizers, pesticides) to farmers, move agricultural products from farms to processing facilities and markets, and deliver food and agricultural commodities to consumers. Supply chain management involves logistics, transportation, storage, and distribution activities to ensure timely delivery and minimize losses.

4.2.5 Food Production

The agricultural industry is crucial for feeding the world's growing population. It produces a wide range of food crops, livestock, and other agrarian products essential for human nutrition and well-being.

4.2.6 Economic Growth

Agriculture contributes significantly to economic growth and development in many countries, providing employment opportunities, generating income for rural communities, and supporting related industries such as food processing, distribution, and retail.

4.2.7 Ecosystem Services

Well-managed agricultural systems can provide important ecosystem services, such as carbon sequestration, soil fertility, water regulation, and biodiversity conservation. Sustainable agriculture practices, such as crop rotation, agroforestry, and integrated pest management, promote soil health and ecosystem resilience.

4.2.8 Technology and Innovation

The agriculture industry is continuously evolving, driven by technological advancements, scientific research, and innovation. Modern agricultural practices incorporate precision farming technologies, genetically modified crops, bioengineering techniques, and digital agriculture tools to optimize resource use, increase productivity, and improve sustainability.

4.2.9 Environmental Sustainability

Sustainable agriculture practices aim to curtail environmental impacts, conserve natural resources, and promote ecosystem health and resilience. Sustainable agriculture techniques include conservation tillage, crop rotation, cover cropping, IPM, agroforestry, and organic farming methods.

4.2.10   Policy and Regulation

Government policies and regulations play a critical role in shaping the agriculture industry, including agricultural subsidies, trade policies, environmental regulations, food safety standards, and land-use planning. Policy decisions influence farm income, market prices, land management practices, and food security outcomes.

4.2.11   Environmental Degradation and Resource Depletion

Intensive agricultural practices can lead to environmental degradation, including soil erosion, water pollution, habitat destruction, and loss of biodiversity. Agriculture is a major consumer of natural resources, including land, water, and energy. Unsustainable agricultural practices, such as monoculture farming, overgrazing, and deforestation, can lead to soil degradation, water scarcity, and depletion of natural habitats.

4.2.12   Climate Change

Agriculture is both a culprit and a victim of climate change. Greenhouse gas emissions from agricultural activities, such as livestock production and deforestation, contribute to global warming. Climate change also affects agricultural productivity, crop yields, and food security, leading to increased risks of crop failures, food shortages, and economic losses.

4.2.13   Challenges and Opportunities

The agriculture industry faces various challenges, including climate change, soil degradation, water scarcity, pest and disease outbreaks, market volatility, and socioeconomic disparities. However, it also presents opportunities for innovation, entrepreneurship, rural development, and sustainable food production to meet the needs of a growing global population.

Overall, the agriculture industry is a diverse and dynamic sector that plays a fundamental role in ensuring food security, supporting economic development, and fostering environmental stewardship on a global scale. Balancing the demands of food production with environmental sustainability and social equity is essential for building a resilient and inclusive agriculture system for future generations.

4.3   Agricultural Industry as a Blessing for Soil Organisms

Agriculture can bring several blessings to soil organisms when managed sustainably and with care (Tahat et al. 2020). Some of the positive impacts of agriculture on soil organisms are as follows:

4.3.1   Organic Matter Input

Agriculture often involves the addition of organic matter to the soil in the form of crop residues, compost, manure, and cover crops. This organic matter serves as a food source for soil organisms, including bacteria, fungi, earthworms, and other decomposers, promoting their growth and activity (Sahu et al. 2019).

4.3.2   Soil Aggregation

Agricultural practices such as reduced tillage, cover cropping, and crop rotation can improve soil structure and aggregation. Soil organisms, particularly earthworms and soil-dwelling arthropods, play essential roles in soil aggregation by burrowing, ingesting organic matter, and excreting castings, which bind soil particles together and improve soil porosity.

4.3.3   Nutrient Cycling

Soil organisms play strategic part in nutrient cycling processes, such as decomposition, mineralization, and nitrogen fixation. Agriculture supports diverse soil microbial communities that contribute to nutrient cycling by decomposing organic matter, releasing nutrients in forms available to plants, and fixing atmospheric nitrogen into plant-available forms.

4.3.4   Biological Pest Control

Agriculture can harness the natural pest control services provided by soil organisms, including predatory insects, nematodes, and microorganisms. IPM strategies promote the conservation of natural enemies of pests, reducing the need for chemical pesticides and enhancing ecological balance in agroecosystems.

4.3.5   Pollination Services

Soil organisms, such as ground-nesting bees, ants, and other insects, contribute to pollination services in agricultural landscapes. Healthy soil habitats provide nesting sites and food resources for pollinators, supporting crop pollination and fruit set, which are essential for agricultural productivity and food security.

4.3.6   Soil Fertility

Agriculture relies on soil fertility for crop production, and soil organisms play critical roles in maintaining soil fertility and productivity, reducing the need for external inputs such as synthetic fertilizers.

4.3.7   Carbon Sequestration

Soil organisms contribute to carbon sequestration in agricultural soils through the incorporation of organic matter into the soil and the formation of stable organic carbon pools. Sustainable agricultural practices that promote soil health and biodiversity can enhance soil carbon storage and mitigate climate change.

4.3.8   Biodiversity Conservation

Agricultural landscapes can support diverse soil organism communities, contributing to overall biodiversity conservation. Well-managed agricultural systems that incorporate agroecological principles, such as crop diversification, agroforestry, and habitat restoration, provide habitat and resources for a wide range of soil organisms.

Thus, agriculture can provide several blessings to soil organisms when practiced sustainably and with consideration for soil health and biodiversity conservation. By promoting soil fertility, nutrient cycling, pest control, and biodiversity conservation, agriculture can support resilient and productive agroecosystems that benefit both humans and the environment.

4.3.9   Agricultural Industry as a Curse for Soil Organisms

While the agricultural industry is essential for food production and livelihoods, it can indeed have negative impacts on soil organisms and soil health if not managed sustainably. Following are the averments on the basis of which agriculture can be perceived as a curse for soil organisms:

4.3.10   Nitrogen Fixation

Synthetic fertilizers, particularly nitrogen-based fertilizers, can affect nitrogen-fixing bacteria such as *Rhizobia* and *Azotobacter* (Abd-Alla et al. 2023). High nitrogen levels in soil can suppress the activity of nitrogen-fixing bacteria, reducing biological nitrogen fixation rates and increasing reliance on external nitrogen inputs. Overreliance on synthetic nitrogen fertilizers can also lead to soil acidification, nutrient imbalances, and loss of soil organic matter, affecting soil health and long-term fertility.

4.3.11   Soil Fauna

Soil-dwelling invertebrates, including earthworms, insects, and microarthropods, are sensitive to agrochemicals such as pesticides and herbicides. Many pesticides are designed to target pest insects but can also harm beneficial insects and soil-dwelling organisms. Earthworm populations, for example, may decline in response to pesticide exposure, disrupting soil structure, organic matter decomposition, and nutrient cycling processes. Predatory and parasitic insects that help control pest populations may also be affected by pesticide residues (Beaumelle et al. 2023).

4.3.12   Plant-Microbe Interactions

Agrochemicals can disrupt beneficial plant-microbe interactions, such as mycorrhizal associations and rhizosphere interactions. Mycorrhizal fungi form symbiotic relationships with plant roots, enhancing nutrient uptake and plant resilience to environmental stresses. Pesticides and herbicides can inhibit mycorrhizal fungi growth and activity, reducing their ability to form associations with plant roots and support plant growth and health. This can lead to decreased nutrient uptake efficiency and increased susceptibility to diseases and environmental stresses (Pagano et al. 2023).

4.3.13   Microbial Communities

Soil microorganisms, including bacteria, fungi, and archaea, play crucial roles in nutrient cycling, decomposition, and soil fertility. Excessive fertilizer application can lead to shifts in microbial community structure, favoring certain microbial groups over others and reducing microbial diversity (Dincă et al. 2022). Pesticides can also have direct toxic effects on soil microbes, inhibiting their growth and activity.

4.3.14   Soil Resilience

Continuous or excessive use of agrochemicals can degrade soil quality and resilience over time, leading to long-term environmental consequences. Soil organisms play critical roles in maintaining soil health and ecosystem resilience, and disruptions to soil communities can have cascading effects on ecosystem functioning. IPM strategies, organic farming practices, and agroecological approaches aim to minimize reliance on agrochemical inputs and promote sustainable soil management practices that support soil biodiversity and ecosystem services.

4.3.15   Social Inequity

The benefits of the agricultural industry are not evenly distributed, leading to social inequities and disparities. Smallholder farmers, indigenous communities, and rural populations often face challenges such as land tenure insecurity, lack of access to resources, and limited market opportunities. Industrialized agriculture can also lead to concentration of land ownership, displacement of small farmers, and loss of traditional farming practices.

In conclusion, the impacts of agrochemicals on soil organisms can vary depending on the type of chemical, application rate, timing, and soil conditions. Sustainable soil management practices that prioritize soil health, biodiversity conservation, and ecosystem resilience are essential for mitigating the negative effects of agrochemicals on soil organisms and promoting long-term agricultural sustainability.

References

Abd-Alla MH, Al-Amri SM, El-Enany AWE (2023) Enhancing *rhizobium*–legume symbiosis and reducing nitrogen fertilizer use are potential options for mitigating climate change. Agriculture 13:2092. https://doi.org/10.3390/agriculture13112092

Beaumelle L, Tison L, Eisenhauer N, Hines J, Malladi S, Pelosi C, Thouvenot L, Phillips HRP (2023) Pesticide effects on soil fauna communities-a meta-analysis. J Appl Ecol 60:1239–1253. https://doi.org/10.1111/13652664.14437

Dincă LC, Grenni P, Onet C, Onet A (2022) Fertilization and soil microbial community: a review. Appl Sci 12:1198. https://doi.org/10.3390/app12031198

Pagano MC, Kyriakides M, Kuyper TW (2023) Effects of pesticides on the arbuscular mycorrhizal symbiosis. Agrochemicals 2:337–354. https://doi.org/10.3390/agrochemicals2020020

Pawlak K, Kołodziejczak M (2020) The role of agriculture in ensuring food security in developing countries: considerations in the context of the problem of sustainable food production. Sustain For 12:5488. https://doi.org/10.3390/su12135488

Sahu P, Singh D, Prabha R, Meena K, Abhilash P (2019) Connecting microbial capabilities with the soil and plant health: options for agricultural sustainability. Ecol Indic 105:601–612

Tahat MM, Alananbeh KA, Othman YI, Leskovar D (2020) Soil health and sustainable agriculture. *Sustainability* 12:4859. https://doi.org/10.3390/su12124859

Chapter 5
Bioindicators: *The Eco-sensors for Detecting Soil Pollution*

Abstract Bioindicators are vital tools for monitoring soil pollution, helping detect, assess, and mitigate contaminants from sources like industrial activities, agriculture, waste disposal, and urbanization. These indicators, including plants, soil microorganisms, earthworms, soil invertebrates, and biochemical markers, respond sensitively to changes in soil quality caused by pollutants. For instance, certain plants can accumulate pollutants, visibly indicating contamination levels, while soil microorganisms and invertebrates reflect changes in community structure, abundance, and behavior under pollution stress. Biochemical markers further provide direct evidence of soil pollution through changes in enzyme activity related to pollutant degradation or detoxification. This chapter discusses the importance of studying these bioindicators to monitor and maintain soil health.

5.1 Introduction

The diversity of soil organisms, particularly microorganisms, is crucial for soil functioning and ecosystem health. Microbes, including bacteria and fungi, play key roles in processes such as decomposition and nutrient cycling. Despite their small size, microorganisms have a momentous impact on soil properties and ecosystem dynamics.

Bioindicators serve as invaluable tools for monitoring soil pollution, enabling us to detect, assess, and mitigate the presence of contaminants in the soil (Bhaduri et al. 2018). Soil pollution, stemming from diverse sources like industrial activities, agricultural practices, waste disposal, and urbanization, poses significant threats to human health, ecosystem integrity, and agricultural sustainability (Dağhan and Ozturk 2015). With the help of bioindicators, researchers can effectively gauge the health of soil ecosystems and the extent of contamination. These indicators, which include various organisms such as plants, soil microorganisms, earthworms, and soil invertebrates, as well as biochemical markers, respond sensitively to changes in soil quality induced by pollutants. For instance, certain plant species have the remarkable ability to accumulate pollutants, offering visible signs of contamination levels. Soil microorganisms, vital for nutrient cycling and soil health, exhibit alterations in community structure and function in response to pollution stressors. Similarly, earthworms and other soil invertebrates serve as indicators of soil health by showing changes in abundance, diversity, and behavior under polluted conditions. Moreover, biochemical markers provide direct evidence of soil pollution by reflecting changes in enzyme activity associated with pollutant degradation or detoxification pathways. Thus, it is imperative to study indicators in soil which provide the signaling system regarding the soil health. The various biological indicators in soil are discussed in this chapter.

5.2   Indicators in Soil

5.2.1   Plant Indicators

Certain plant species, known as bioindicator plants, are sensitive to specific soil conditions, pollutants, or environmental stressors. Changes in plant species composition, diversity, or health can indicate soil characteristics such as pH, nutrient levels, pollution levels, or habitat quality. Plant indicators are commonly used in ecological assessments, restoration projects, and soil monitoring programs.

5.2.2   Chemical Indicators

Soil chemical properties such as pH, nutrient levels, organic matter content, electrical conductivity, and heavy metal concentrations can serve as indicators of soil health, fertility, and pollution. Monitoring changes in soil chemistry over time can help assess soil conditions, nutrient availability, contamination levels, and the effectiveness of soil management practices.

5.2.3   Physical Indicators

Soil physical properties such as texture, structure, compaction, porosity, and water-holding capacity influence soil fertility, water infiltration, root growth, and erosion susceptibility. Changes in soil physical properties can indicate alterations in soil health, structure stability, and erosion risk. Physical indicators are often used to assess soil erosion, compaction, and habitat suitability for soil organisms.

5.2.4 Functional Indicators

Soil functional indicators focus on specific soil processes and functions, such as nutrient cycling, carbon sequestration, water filtration, and pest regulation. Monitoring changes in functional indicators can provide insights into soil health, ecosystem functioning, and the delivery of ecosystem services. Functional indicators are often used in ecological assessments, land management practices, and ecosystem restoration projects.

5.2.5 Bioindicators in Soil

Bioindicators are organisms or biological processes that provide information about the quality of the habitat they inhabit. In soil quality assessment, bioindicators play a crucial role in indicating the health and overall condition of the soil ecosystem. These indicators can reflect various aspects of soil quality, including its fertility, contamination levels, and overall ecological function. Few types of soil bioindicators are discussed below:

5.2.6 Microbial Indicators

Soil microbes, including bacteria, fungi, archaea, and protozoa, play essential roles in nutrient cycling, organic matter decomposition, and soil fertility. Microbial indicators such as microbial biomass, microbial diversity, enzyme activities, and microbial community composition can provide insights into soil health, microbial activity, and ecosystem functioning.

5.2.7 Soil Enzymes as Bioindicators

In the context of environmental analysis tools, assessing soil enzymes exemplifies one of the most sensitive and reliable approaches for evaluating soil health. Enzymes serve as bioindicators in soils affected by different environmental factors, including xenobiotics such as heavy metals, stress conditions, and management practices. They play essential roles in catalyzing biochemical reactions necessary for microbial and plant growth, nutrient cycling, organic matter decomposition, gas exchange, and aggregate formation.

Soil enzymes can be classified into four main groups:

 (i) *Oxidoreductases*: These enzymes catalyze oxidation-reduction reactions.
 (ii) *Transferases*: Involved in transferring functional groups between molecules.

(iii) *Hydrolases*: Catalyze the hydrolysis of various organic compounds.
(iv) *Lyases*: Responsible for eliminating groups from molecules to form double bonds.

Among these groups, oxidoreductases and hydrolases are particularly important in soil biology. Oxidoreductases are involved in redox processes, while hydrolases play critical roles in the breakdown of complex organic compounds. Measuring soil enzyme activities provides valuable insights into soil functionality, health, and responses to environmental changes. Enzyme activity assays can be entitled as indicators of soil quality, pollution levels, and ecosystem resilience. They offer a rapid and sensitive means of assessing soil health and identifying potential environmental risks, particularly in agricultural soils subjected to intensive management practices or contamination with pollutants like heavy metals.

5.2.8 Mesofauna Indicators

Soil mesofauna, including mites, springtails, nematodes, and collembolans, are small invertebrates that play strategic activities in nutrient cycling, soil structure maintenance, and decomposition processes. Changes in mesofauna abundance, diversity, and activity can indicate soil health, microbial activity, and organic matter decomposition rates. Mesofauna indicators are often used to assess soil biodiversity and ecosystem functioning. Collembolans are sensitive to desiccation/drought as they require moisture-enriched habitat in order to meet the needs of their tegumentary respiration. Some springtail species, such as *Folsomia candida* and *Tomocerus minor*, are sensitive to heavy metals and pesticides (Dervash et al. 2018), whereas *Galumna obvia* (Fig. 5.1) has emerged as a pollution-resilient mite (Minodora et al. 2019).

Fig. 5.1 Stereomicroscopic view of *Galumna obvia*

5.2.9 Macrofauna Indicators

Soil macrofauna, including earthworms, ants, termites, beetles, and millipedes, are important decomposers, predators, and soil engineers. Changes in macrofauna abundance, diversity, and community composition can reflect alterations in soil health, habitat quality, and disturbance levels.

Few types of soil organisms commonly used as bioindicators of soil pollution include the following:

(i) *Earthworms*: Earthworms are key decomposers and ecosystem engineers in soil ecosystems. Certain species, such as *Eisenia fetida* and *Lumbricus terrestris*, are sensitive to soil pollutants like heavy metals and pesticides. Changes in earthworm abundance, diversity, and behavior can indicate soil pollution levels and impacts on soil health (Hirano and Tamae 2011).

(ii) *Ants*: Ants play imperative roles in soil nutrient cycling, seed dispersal, and soil structure modification. Some ant species, such as *Formica* spp. and *Lasius* spp., are sensitive to environmental contaminants. Monitoring ant abundance, species composition, and foraging behavior can provide insights into soil pollution impacts on ant communities and ecosystem processes (Frouz and Jílková 2008).

(iii) *Dung Beetles*: Dung beetles are important detritivores and nutrient recyclers in soil ecosystems, contributing to dung decomposition and soil fertility. Certain dung beetle species, such as *Onthophagus* spp. and *Aphodius* spp., are sensitive to soil pollutants. Changes in dung beetle abundance, diversity, and activity can indicate soil pollution levels and impacts on nutrient cycling and ecosystem function (Ma et al. 2023).

(iv) *Ground Beetles*: Ground beetles are predatory insects that play roles in regulating soil-dwelling invertebrate populations and ecosystem dynamics. Certain ground beetle species, such as *Carabidae* spp. and *Staphylinidae* spp., are sensitive to environmental contaminants. Changes in ground beetle abundance, diversity, and behavior can indicate soil pollution levels and impacts on soil food webs and ecosystem stability (Eyre et al. 2015).

(v) *Termites*: Termites are important decomposers and ecosystem engineers in tropical and subtropical soils, contributing to organic matter decomposition and soil structure formation (Bignell 2006). Some termite species, such as *Macrotermes* spp. and *Coptotermes* spp., are sensitive to soil pollutants. Monitoring termite abundance, diversity, and mound-building activities can provide insights into soil pollution impacts on termite communities and ecosystem processes (Bignell 2006).

(vi) *Millipedes as Bioindicators*: Several species within the Julidae family have been investigated for their potential as bioindicators of soil pollution, particularly in agricultural areas where pesticide use is common. Similarly, *Tachypodoiulus niger* is widespread across Europe and has been studied in the context of organic pollution in soil. This species can bioaccumulate organic pollutants such as polycyclic aromatic hydrocarbons (PAHs) from contaminated soil, providing a measurable indicator of PAH pollution levels in terrestrial ecosystems. *Glyphiulus granulatus* is a small millipede species that has been investigated as a bioindicator of soil pollution, particularly in areas affected by industrial activities or urbanization. Studies have shown that *Glyphiulus granulatus* can exhibit changes in population abundance and diversity in response to soil pollution, making it useful for assessing the ecological impacts of contamination (Schapheer et al. 2021). Research has shown that *Orthoporus ornatus* and *Spirobolida* millipedes can accumulate heavy metals such as lead, cadmium, and zinc in its body tissues when exposed to contaminated soil, making it useful for assessing metal pollution levels in terrestrial environments (Sridhar 2011, Rost-Roszkowska et al. 2021).

(vii) *Caterpillars as bioindicators*: The Elephant Hawk-moth (*Deilephila elpenor*) is a notable caterpillar species within the Sphingidae family and can serve as a bioindicator due to its specific habitat requirements and sensitivity to environmental changes. As a bioindicator, it can provide valuable insights into the health of ecosystems, particularly in relation to habitat quality and the presence of certain plant species. Elephant Hawk-moths are primarily found in habitats rich in their larval food plants, such as bedstraw and willow herb. Their presence indicates a healthy environment with a diverse plant community, which is essential for their survival. Moreover, changes in their population dynamics can signal shifts in environmental conditions, such as pollution levels, climate change, and habitat destruction (Dar and Jamal 2021).

Similarly, the presence and health of vine hawkmoth caterpillar (*Hippotion celerio*) populations can indicate soil contamination levels (Fig. 5.2). Pollutants and toxins in the soil can accumulate in plants and affect herbivorous insects like the *Hippotion celerio* larvae (family Choerocampinae), leading to observable changes in their growth, development, and survival rates. Monitoring these changes can provide valuable insights into the presence of harmful substances in the soil (Dar and Jamal 2021).

(viii) *Spiders as Bioindicators*: Spiders are particularly useful as bioindicators due to their sensitivity to environmental disturbances, wide distribution, and the relatively well-understood nature of their ecology and life history (Holt and Miller 2010). The presence and health of huntsman spiders (family Sparassidae) can indicate the level of soil contamination, the abundance of prey species, and overall ecosystem health. Since huntsman spiders are common and relatively easy to survey, they meet several criteria for effective bioindicators, including being abundant, having a stable taxonomy, and being well-studied (Holt and Miller 2010). Furthermore, their responses to environmental stressors are typically measurable, providing valuable data for ecological assessments.

Pholcus phalangioides, commonly known as the daddy long legs spider (depicted in Fig. 5.3), has emerged as a potential bioindicator of soil health in

Fig. 5.2 The vine hawkmoth (*Hippotion celerio*) populations

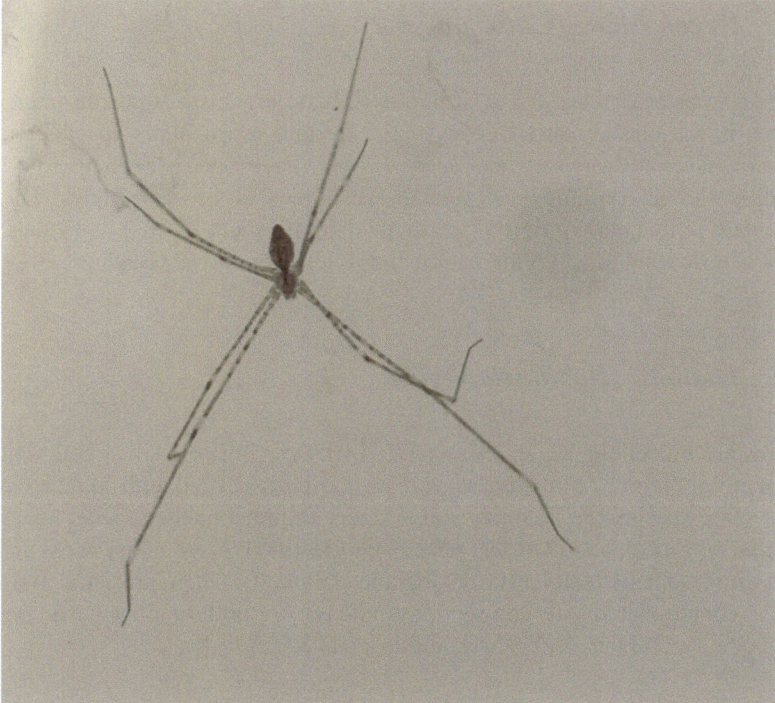

Fig. 5.3 The daddy long legs spider (*Pholcus phalangioides*)

agroecosystems due to its sensitivity to environmental changes and its close association with soil conditions. *Pholcus phalangioides* plays a crucial role in regulating insect populations and nutrient cycling within soil ecosystems (Nyffeler and Sunderland 2003). Studies have shown a positive correlation between the presence of *Pholcus phalangioides* and soil health indicators such as soil organic carbon, microbial biomass, and soil structure stability (Ghavami 2007). Understanding the ecological role of *Pholcus phalangioides* as a bioindicator species provides valuable insights into the overall health and functioning of soil ecosystems.

5.3 Algae as Bioindicator

While algae are primarily associated with aquatic environments, certain algal species can indeed inhabit soil surfaces and play important roles in soil ecosystems. Some algae have developed mechanisms to tolerate or degrade pollutants, making them valuable bioindicators of soil pollution. Types of algae that serve as bioindicators include the following:

5.3.1 Green Algae (Chlorophyta)

Green algae are a diverse group of photosynthetic organisms that can inhabit soil surfaces in various terrestrial ecosystems. Certain green algae species, such as *Chlorella*, *Chlamydomonas*, and *Scenedesmus*, have been shown to tolerate or degrade pollutants such as heavy metals, pesticides, and hydrocarbons. They can accumulate pollutants within their cells or detoxify them through enzymatic processes, contributing to soil remediation and pollutant removal (Baghour 2019).

5.3.2 Diatoms (Bacillariophyta)

Diatoms are unicellular algae characterized by their intricate silica cell walls and diverse morphologies. While diatoms are primarily associated with aquatic environments, some species can colonize soil surfaces in moist habitats. Diatoms such as *Navicula*, *Nitzschia*, and *Diatoma* have been identified in soil ecosystems and may possess mechanisms to tolerate or degrade pollutants. Their presence and abundance in contaminated soils can serve as indicators of pollution levels and environmental quality (Zelazna-Wieczorek and Bogusz 2022).

5.3.3   Blue-Green Algae (Cyanobacteria)

Cyanobacteria, also known as blue-green algae, are photosynthetic bacteria capable of inhabiting diverse habitats, including soil surfaces. Certain cyanobacteria species, such as *Nostoc*, *Anabaena*, and *Microcystis*, have been shown to tolerate or degrade pollutants such as heavy metals, pesticides, and organic contaminants. They can utilize pollutants as energy sources or sequester them within specialized cellular structures, contributing to soil detoxification and remediation (Srivastava et al. 2021).

5.3.4   Filamentous Algae (Charophyta)

Filamentous algae are multicellular algae characterized by their long, threadlike structures. While filamentous algae are predominantly found in aquatic environments, some species can colonize soil surfaces in wetland habitats. Filamentous algae such as *Spirogyra*, *Oedogonium*, and *Cladophora* have been identified in soil ecosystems and may possess mechanisms to tolerate or degrade pollutants (Liu et al. 2023). Their presence and abundance in contaminated soils can indicate pollution levels and ecosystem health.

5.3.5   Eukaryotic Algae

Various eukaryotic algae belonging to different taxonomic groups, including green algae, red algae, and brown algae, can inhabit soil surfaces in terrestrial ecosystems. These algae species may possess mechanisms to tolerate or degrade pollutants, although their specific responses to soil pollution require further research. Their presence and activity in contaminated soils can serve as indicators of pollution levels and ecosystem resilience (Holzinger and Karsten 2013).

5.4   Lichens as Bioindicators

Lichens, with their sensitivity to environmental conditions and pollutants, are frequently employed as bioindicators of soil pollution. Their ability to accumulate pollutants from the atmosphere and soil and their wide distribution and diverse species composition make them valuable tools for assessing soil contamination (Gautam et al. 2022). Some types of lichens commonly used as bioindicators of soil pollution include the following:

5.4.1   Sensitive Species

Certain lichen species are particularly sensitive to pollutants and demonstrate visible changes in their growth, morphology, or physiology in polluted environments. These species are often used as indicators of soil pollution and can provide early warnings of contamination. Examples of sensitive lichen species include *Lecanora conizaeoides*, *Flavoparmelia caperata*, and *Parmelia sulcata* (Yang et al. 2023).

5.4.2   Accumulative Species

Some lichen species have the ability to accumulate pollutants from the atmosphere and soil over time, resulting in high concentrations of contaminants within their tissues. These species act as bioaccumulators of pollutants and can be used to assess the extent and severity of soil pollution. Examples of accumulative lichen species include *Hypogymnia physodes*, *Usnea* spp., and *Xanthoria* spp. (Białońska and Dayan 2005).

5.4.3   Indicator Species

Certain lichen species exhibit specific responses to particular pollutants or environmental stressors, making them valuable indicators of soil pollution. Changes in the abundance, diversity, or distribution of these indicator species can indicate variations in soil pollution levels and types of contaminants (Frati and Brunialti 2023). Examples of indicator lichen species include *Stereocaulon* spp., which are sensitive to heavy metals, and *Ramalina* spp., which are sensitive to sulfur dioxide.

5.4.4   Community Composition

Lichen community composition and diversity can reflect soil pollution gradients and contamination sources. Monitoring changes in lichen community structure, species richness, and dominance patterns can provide insights into soil pollution levels, pollutant sources, and environmental impacts. Shifts in lichen communities toward pollution-tolerant or pollution-sensitive species can indicate changes in soil quality and contamination severity (Frati and Brunialti 2023).

5.4.5 Transplant Experiments

Transplant experiments involving the exposure of lichen thalli to contaminated soils or atmospheric deposition can provide direct evidence of lichen responses to soil pollution. By transplanting lichen samples from clean sites to polluted sites or vice versa, researchers can assess lichen tolerance, bioaccumulation rates, and pollutant uptake mechanisms. Transplant experiments can help validate lichen bioindicators and quantify pollutant levels in soil environments (Yang et al. 2023).

Overall, lichens and their various types serve as valuable bioindicators of soil pollution, offering insights into contamination levels, pollutant types, and environmental impacts. By monitoring lichen populations and communities in soil environments, researchers, land managers, and policymakers can assess soil pollution, diagnose environmental stressors, and implement remediation measures to protect soil health and ecosystem integrity.

5.5 Microbes as Bioindicators

The study of soil microorganisms is crucial for understanding soil health, as they play vital roles in nutrient cycling, decomposition of organic matter, and overall soil fertility. Various techniques are employed to analyze soil microbial communities, ranging from culturable methods to molecular techniques based on genomic approaches. However, molecular techniques are often preferred due to their ability to identify nonculturable microorganisms and provide a more comprehensive understanding of microbial diversity (Aguilar-Paredes et al. 2023).

5.6 Archaea

5.6.1 Methanogens

Methanogenic archaea are anaerobic microorganisms that produce methane as a metabolic byproduct. Changes in the abundance or activity of methanogens can indicate variations in soil pollution levels, particularly in anaerobic environments contaminated with organic pollutants. Monitoring methane production rates or methanogen populations can provide insights into soil pollution impacts and remediation potential (Domínguez-Espinosa et al. 2023).

5.6.2   Ammonia-Oxidizing Archaea (AOA)

AOA are archaea capable of oxidizing ammonia to nitrite, playing important roles in soil nitrogen cycling. Changes in AOA abundance or diversity can indicate variations in soil pollution levels, particularly in soils contaminated with nitrogenous pollutants such as ammonia or nitrate (Domínguez-Espinosa et al. 2023). Monitoring AOA populations can provide insights into soil pollution impacts on nitrogen dynamics and ecosystem functioning.

5.6.3   Bacterial Bioindicators

Certain bacterial species are known to be sensitive to specific pollutants, such as heavy metals, pesticides, hydrocarbons, and organic contaminants. Changes in the abundance, diversity, or composition of bacterial communities can indicate variations in soil pollution levels and types of contaminants (Sazykin et al. 2023). For example, metal-tolerant bacteria like *Pseudomonas* species are used as indicators of heavy metal pollution.

5.7   Protozoa as Bioindicators

Certain protozoan species are sensitive to pollutants such as heavy metals, pesticides, and organic contaminants. Changes in protozoan community structure or abundance can indicate variations in soil pollution levels and ecosystem health (Yuan et al. 2024). For example, changes in the abundance of protozoan predators like *amoebae* and *flagellates* can indicate shifts in soil pollution impacts on microbial communities and ecosystem functioning.

5.8   Fungal Bioindicators in Soil

Some fungal species possess incredible capabilities to tolerate or degrade various pollutants, making them invaluable bioindicators of soil pollution. These fungi have evolved mechanisms to detoxify or utilize pollutants as energy sources, contributing to the natural remediation of contaminated environments (Vaksmaa et al. 2023):

5.8.1 White-Rot Fungi

White-rot fungi are renowned for their capacity to degrade complex organic pollutants, including polycyclic aromatic hydrocarbons (PAHs), polychlorinated biphenyls (PCBs), pesticides, and dyes. Species such as *Phanerochaete chrysosporium*, *Trametes versicolor*, and *Pleurotus ostreatus* produce ligninolytic enzymes like lignin peroxidase, manganese peroxidase, and laccase, which break down recalcitrant organic compounds. Their activity can serve as an indicator of soil pollution and remediation potential (Latif et al. 2023).

5.8.2 Arbuscular Mycorrhizal Fungi (AMF)

AMF form symbiotic associations with plant roots, enhancing nutrient uptake and plant growth. Some AMF species, such as *Rhizophagus irregularis* (formerly *Glomus intraradices*), have been shown to promote the degradation of pollutants like petroleum hydrocarbons and heavy metals by enhancing plant tolerance and stimulating microbial activity in the rhizosphere. Their presence and activity can indicate soil pollution levels and remediation potential in contaminated sites (Deja-Sikora et al. 2023).

5.8.3 Endophytic Fungi

Endophytic fungi reside inside plant tissues without causing harm to the host and can contribute to plant health and stress tolerance. Certain endophytic fungi, such as *Cladosporium*, *Penicillium*, and *Aspergillus* species, possess the ability to degrade pollutants such as hydrocarbons, pesticides, and herbicides. They can be isolated from plants growing in contaminated soils and used as bioindicators of soil pollution and phytoremediation potential (Fadiji and Babalola 2020).

5.8.4 Ascomycetes and Basidiomycetes

Various ascomycete and basidiomycete fungi have been identified as pollutant degraders in soil ecosystems. For example, certain species of *Aspergillus*, *Fusarium*, *Penicillium*, *Streptomyces*, and *Cladosporium* can degrade a wide range of organic pollutants, including aromatic compounds, herbicides, and insecticides. Their presence and activity in contaminated soils can indicate pollutant degradation potential and ecosystem resilience (Matúš et al. 2023).

5.8.5   Metal-Tolerant Fungi

Some fungal species exhibit high tolerance to heavy metals and can thrive in metal-contaminated soils. Metal-tolerant fungi such as *Paxillus involutus, Suillus luteus*, and *Rhizopus arrhizus* can accumulate heavy metals within their mycelia or sequester them in intracellular compartments, reducing metal toxicity and facilitating soil phytoremediation. Their abundance and diversity can serve as indicators of heavy metal pollution and remediation potential (Amir et al. 2014).

These fungal species, along with their pollutant-degrading capabilities, can be utilized as bioindicators of soil pollution in contaminated environments. By monitoring their presence, abundance, and activity, researchers can assess soil pollution levels, track pollutant degradation processes, and implement targeted remediation strategies to restore soil health and ecosystem function.

By using a combination of these soil bioindicators, researchers, land managers, and policymakers can assess soil health, diagnose soil degradation, and implement sustainable land management practices to maintain or restore soil ecosystems' productivity and resilience.

References

Aguilar-Paredes A, Valdés G, Araneda N, Valdebenito E, Hansen F, Nuti M (2023) Microbial community in the composting process and its positive impact on the soil biota in sustainable agriculture. Agronomy 13:542. https://doi.org/10.3390/agronomy13020542

Amir H, Jourand P, Cavaloc Y, Ducousso M (2014) Role of mycorrhizal fungi in the alleviation of heavy metal toxicity in plants. In: Mycorrhizal fungi: use in sustainable agriculture and land restoration. https://doi.org/10.1007/978-3-662-45370-4_15

Baghour M (2019) Algal degradation of organic pollutants. In: Handbook of ecomaterials, pp 565–586. https://doi.org/10.1007/978-3-319-68255-6_86

Bhaduri D, Chatterjee D, Chakraborty K, Chatterjee S, Saha A (2018) Bioindicators of degraded soils. In: Lichtfouse E et al (eds) Sustainable agriculture reviews. Springer, Cham, pp 231–257. https://doi.org/10.1007/978-3-319-99076-7_8

Białońska D, Dayan F (2005) Chemistry of the lichen *Hypogymnia physodes* transplanted to an industrial region. J Chem Ecol 31:2975–2991. https://doi.org/10.1007/s10886-005-8408-x

Bignell D (2006) Termites as soil engineers and soil processors. In: Intestinal microorganisms of termites and other invertebrates. Springer, Berlin, pp 183–220. https://doi.org/10.1007/3-540-28185-1_8

Dağhan H, Ozturk M (2015) Soil pollution in Turkey and remediation methods. In: Hakeem KR, Sabir M, Ozturk M, Mermut AR (eds) Soil remediation and plants: prospects and challenges, pp 287–312

Dar A, Jamal K (2021) Moths as ecological indicators: a review. Munis Entomol Zool J 16:830–836

Deja-Sikora E, Werner K, Hrynkiewicz K (2023) AMF species do matter: *rhizophagus irregularis* and *Funneliformis mosseae* affect healthy and PVY-infected *Solanum tuberosum* L. in a different way. Front Microbiol 14:1127278. https://doi.org/10.3389/fmicb.2023.1127278

Dervash MA, Bhat RA, Mushtaq N, Singh DV (2018) Dynamics and importance of soil Mesofauna. Int J Adv Res Sci Eng 7:2010–2019

Domínguez-Espinosa ME, Cruz-Salomón A, Ramírez de León JA, Hernández-Méndez JME, Santiago-Martínez MG (2023) Syntrophy between bacteria and archaea enhances methane

production in an EGSB bioreactor fed by cheese whey wastewater. Front Sustain Food Syst 7:1244691. https://doi.org/10.3389/fsufs.2023.1244691

Eyre M, McMillan SD, Critchley N (2015) Ground beetles (Coleoptera, Carabidae) as indicators of change and pattern in the agroecosystem: longer surveys improve understanding. Ecol Indic 68. https://doi.org/10.1016/j.ecolind.2015.11.009

Fadiji AE, Babalola OO (2020) Elucidating mechanisms of endophytes used in plant protection and other bioactivities with multifunctional prospects. Front Bioeng Biotechnol 8:467. https://doi.org/10.3389/fbioe.2020.00467

Frati L, Brunialti G (2023) Recent trends and future challenges for lichen biomonitoring in forests. Forests 14:647. https://doi.org/10.3390/f14030647

Frouz J, Jílková V (2008) The effect of ants on soil properties and processes (hymenoptera: Formicidae). Myrmecol News 11:191–199

Gautam M, Mishra S, Agrawal M (2022) Bio-indicators of soil contaminated with organic and inorganic pollutants. In: Tiwari S, Agrawal SB (eds) New paradigms in environmental biomonitoring using plants. Elsevier, pp 271–298. https://doi.org/10.1016/B978-0-12-824351-0.00001-8

Ghavami S (2007) Spider fauna in Caspian costal region of Iran. Pak J Biol Sci 10(5):682–691. https://doi.org/10.3923/pjbs.2007.682.691

Hirano T, Tamae K (2011) Earthworms and soil pollutants. Sensors (Basel) 11:11157–11167. https://doi.org/10.3390/s111211157

Holt EA, Miller SW (2010) Bioindicators: using organisms to measure environmental impacts. Nat Educ Knowl 3(10):8

Holzinger A, Karsten U (2013) Desiccation stress and tolerance in green algae: consequences for ultrastructure, physiological and molecular mechanisms. Front Plant Sci 4:327. https://doi.org/10.3389/fpls.2013.00327

Latif W, Ciniglia C, Iovinella M, Shafiq M, Papa S (2023) Role of white rot fungi in industrial wastewater treatment: a review. Appl Sci 13:8318. https://doi.org/10.3390/app13148318

Liu J, Pemberton B, Scales PJ, Martin GJO (2023) Ammonia tolerance of filamentous algae Oedogonium, Spirogyra, Tribonema and Cladophora, and its implications on wastewater treatment processes. Algal Res. https://doi.org/10.1016/j.algal.2023.103126

Ma L, Weeraratne N, Gurusinghe S, Aktar J, Haque KMS, Eberbach P, Gurr GG, Weston LA (2023) Dung beetle activity is soil-type-dependent and modulates pasture growth and associated soil microbiome. Agronomy 13:325. https://doi.org/10.3390/agronomy13020325

Matúš P, Littera P, Farkas B, Urík M (2023) Review on performance of *Aspergillus* and *Penicillium* species in biodegradation of organochlorine and organophosphorus pesticides. Microorganisms 11(6):1485. https://doi.org/10.3390/microorganisms11061485

Minodora M, Honciuc V, Neagoe A, Băncilă R, Iordache V, Onete M (2019) Soil mite communities (Acari: Mesostigmata, Oribatida) as bioindicators for environmental conditions from polluted soils. Sci Rep 9:20250. https://doi.org/10.1038/s41598-019-56700-8

Nyffeler M, Sunderland K (2003) Composition, abundance and pest control potential of spider communities in agroecosystems: a comparison of European and US studies. Agric Ecosyst Environ 95:579–612. https://doi.org/10.1016/S0167-8809(02)00181-0

Rost-Roszkowska MM, Vilimová J, Tajovský K et al (2021) Structure of the midgut epithelium in four diplopod species: histology, histochemistry and ultrastructure. Arthropod Syst Phylo 79:295–308

Sazykin I, Khmelevtsova L, Azhogina T, Sazykina M (2023) Heavy metals influence on the bacterial community of soils: a review. Agriculture 13:653. https://doi.org/10.3390/agriculture13030653

Schapheer C, Pellens R, Scherson R (2021) Arthropod-microbiota integration: its importance for ecosystem conservation. Front Microbiol 12:702763. https://doi.org/10.3389/fmicb.2021.702763

Sridhar K (2011) Diversity, restoration and conservation of millipedes. In: Biodiversity in India, pp 1–38. https://doi.org/10.13140/RG.2.1.3683.2889

Srivastava R, Kanda T, Yadav S, Mishra R, Atri N (2021) Cyanobacteria in rhizosphere: dynamics, diversity, and Symbiosis. In: Plant, soil and microbes in tropical ecosystems. https://doi.org/10.1007/978-981-16-3364-5_4

Vaksmaa A, Guerrero-Cruz S, Ghosh P, Zeghal E, Hernando-Morales V, Niemann H (2023) Role of fungi in bioremediation of emerging pollutants. Front Mar Sci 10:1070905. https://doi. org/10.3389/fmars.2023.1070905

Yang J, Oh SO, Hur JS (2023) Lichen as bioindicators: assessing their response to heavy metal pollution in their native ecosystem. Mycobiology 51:343–353. https://doi.org/10.1080/1229809 3.2023.2265144

Yuan G, Chen Y, Wang Y, Zhang H, Wang H, Jiang M, Zhang X, Gong Y, Yuan S (2024) Responses of protozoan communities to multiple environmental stresses (warming, eutrophication, and pesticide pollution). Animals 14:1293. https://doi.org/10.3390/ani14091293

Zelazna-Wieczorek J, Bogusz I (2022) Diatoms from inland aquatic and soil habitats as indestructible and nonremovable forensic environmental evidence. J Forensic Sci 67. https://doi. org/10.1111/1556-4029.15017

Chapter 6
Microbial Goldmine and Organic Farming: *A Key to Agricultural Production and Clean Environment*

Abstract Agriculture, an ancient practice originating around 10,000 B.C. in the Middle East, laid the foundation for sustainable food production through organic farming. However, modern agricultural advancements have led to significant pollution from synthetic pesticides and fertilizers, contaminating soil, air, and water, and introducing harmful substances into the food chain. The use of heavy metal-laden irrigation water exacerbates this issue, posing severe risks to human health and ecosystem integrity. To combat these detrimental effects, traditional organic farming practices and sustainable strategies offer a promising blueprint for environmental conservation. This chapter drives out attention towards the role of microbes in emphasizing soil health through microbial activity and how organic agriculture fosters productivity and resilience, paving the way for a more sustainable and regenerative food system.

6.1 Introduction

Agriculture is an imperial concept initiated and developed by neolithic man in Middle East around 10,000 BC. The scope of organic farming grasped the limelight from time immemorial which formed the backbone of sustainable agriculture. But with each passing span of time, modern agricultural advancements have stemmed up the process of agricultural pollution. Usage of synthetic pesticides and fertilizers results in pollution of soil, air, and water which eventually find its entry into the food chain. The heavy metal-contaminated water utilized for irrigation results in circulation of xenobiotics from abiotic to biotic compartments of ecosystems which are detrimental to humans and other life-forms. Therefore, agricultural pollution sharply targets healthy survival of humans, animals, and various life-forms in an ecosystem. Thus, in order to mitigate detrimental effects of modern agricultural advancements, traditional organic farming and its various sustainable strategies provide blueprint for sustainable agriculture and environmental conservation (Ozturk et al. 2022).

Organic agriculture and microbial activity represent a goldmine of sustainable farming practices that prioritize soil health, biodiversity, and environmental stewardship. By harnessing the power of soil microbes, organic farming systems promote agricultural productivity, resilience, and long-term sustainability, offering a pathway toward a more sustainable and regenerative food system. This chapter drives out attention toward the role of microbes in emphasizing soil health through microbial activity and how organic agriculture fosters productivity and resilience, paving the way for a more sustainable and regenerative food system.

6.2   Conventional Farming and Soil Microbes

Conventional farming plays a momentous function in fulfilling the food amenities of an escalating demographic status, which has resulted in ever-mounting dependence on synthetic soil conditioners, viz., "fertilizers and pesticides." "Synthetic fertilizers" are technologically maneuvered entities chiefly composed of known magnitudes of major nutrients, vis-à-vis their unmanaged utilization results in atmospheric and subsurface water contamination besides accelerated eutrophication of inland waters.

The unwarranted use of "synthetic fertilizers and pesticides" has insensitively inflated the environmental quality and caused havoc to food safety and value, which eventually result in health problems in humans and animals (Mahankale 2024). The detrimental impacts produced by noxious byproducts released by synthetic fertilizers including their health impacts are depicted in Table 6.1.

Accordingly, there has been a rising scope in eco-farming and organic farming by victual clients and ecologists as possible substitutes to synthetic entity-based conventional agriculture. Some microbes play a pivotal function in soil (Liebich et al. 2003) by indispensably improving the soil structure, mineralization of organic matter, and making nutrients accessible for plants (Ozturk et al. 2022; Wei et al. 2024).

Organic agriculture and microbial activity form a dynamic duo that holds immense potential for sustainable farming practices. Soil microbes are also capable of metabolizing and degrading a lot of contaminants and pesticides (Pal et al. 2005) and thus are of immense importance for exploiting in biotechnological industry (Wei et al. 2024). Likewise, microbial degradation can result in configuration of supplementary noxious recalcitrant metabolites. Although soil microbial populations have the characteristics of being more flexible and adaptable to the altered environmental conditions, the incorporation of pesticides on crop patches can cause momentous irremediable alterations in target populations. Environmental pressure at species level can cast a considerable brunt on functions of entire landscape ecosystem. Fungicides are known to be toxic to "soil fungi and actinomycetes" which results in causing negative alterations in "microbial community structure" (Stamenković et al. 2018). Contrarily, "bacterial species, such as nitrification bacteria," are extremely susceptible to pesticide activity, for example, "inhibition of

Table 6.1 Noxious byproducts released by synthetic fertilizers including their health impacts

S. no.	Toxicity with the synthetic fertilizer	Impacts on human health
1.	"Contaminated water (agricultural runoff)"	Contaminated water, particularly runoff from agricultural areas, often contains elevated levels of nitrates and nitrites. These compounds can be harmful to human health, particularly because they interfere with the synthesis of hemoglobin, the protein responsible for transporting oxygen in the blood. When individuals are exposed to high levels of nitrates and nitrites through unclean water sources, such as dirty drinking water or recreational aquatic environs, it can lead to a condition called methemoglobinemia. This medical condition occurs when there's an abnormal amount of methemoglobin in the blood, which reduces its ability to carry oxygen effectively. This can result in symptoms such as shortness of breath, fatigue, headaches, and in severe cases cyanosis (a bluish discoloration of the skin)
2.	"Heavy metals"	Heavy metals are indeed a concern when they are present as impurities in fertilizers. These substances can contaminate soil, water, and crops, posing serious risks to human well-being and the environment. Exposure to heavy metals through contaminated fertilizers can lead to a range of health problems, including kidney damage, respiratory issues (such as lung problems), liver damage, and an increased risk of cancer. Accumulation of heavy metals in the body over time leads to long-lasting health conditions and long-term consequences for affected individuals
3.	"Exposure to ammonium nitrate"	Exposure to ammonium nitrate can indeed result in a range of health issues pertaining to the nose, throat, and respiratory system, primarily due to its irritant properties and potential toxicity. Symptoms include nervousness, nausea, vomiting, headache, uncontrolled muscle twitching, fainting, and even death
4.	"Potassium chloride"	It can interfere with nerve impulses and effectively interrupt whole body functioning and chiefly affects the cardiac working. It can cause redness or itching of the skin or eyes, headaches, gastric pain, dizziness, diarrhea with blood, convulsions, and mental disorders
5.	"Cadmium"	Cadmium poisoning can indeed lead to a condition known as "itai-itai" (ouch-ouch; "it hurts, hurts" in Japanese). This debilitating condition is characterized by severe pain in the joints and spine, among other symptoms. "Itai-itai" disease is primarily caused by chronic exposure to high levels of cadmium, typically through contaminated water or food sources
6.	"Cancers"	Synthetic agrochemicals enhance cancer incidences, including brain cancer, lymphoma, prostate cancer, leukemia, colon cancer, and rectal cancer

nitrification by sulfonyl urea herbicides." Likewise, "chlorothalonil and dinitrophenyl fungicides (such as mancozeb, maneb, or zineb) intoxicate nitrification and denitrification bacterial processes." Moreover, various organochlorine pesticides like pentachlorphenol, DDT, and methyl parathion can disrupt the symbiotic association between leguminous plants and *Rhizobium*. These chemicals can interfere with the signaling pathways that enable the formation of symbiosis between the plant roots and the bacteria, ultimately leading to a reduction in nitrogen fixation efficiency. As a result, the availability of nitrogen for plant uptake diminishes, leading to poorer crop production (Potera 2007).

Indeed, the increased dependence on artificial nitrogenous fertilizers has adverse consequences on fertility of soil and sustainability of crop production. These fertilizers, while initially boosting plant growth by providing readily available nitrogen, can lead to long-term soil degradation and reduced fertility (Potera 2007). When the pesticides like benomyl and dimethoate are applied to agricultural fields, they can inadvertently harm mycorrhizal fungi populations (Menendez et al. 1999). These chemicals may directly inhibit the growth and activity of the fungi or indirectly disrupt their symbiotic relationship with plant roots. As a result, the plant's ability to access essential nutrients through mycorrhizal associations becomes compromised. The repercussions of pesticide-induced damage to mycorrhizal fungi extend beyond individual plants. Since mycorrhizal fungi facilitates nutrient cycling and soil structure, thus, their decline can have far-reaching impacts on soil health and ecosystem functioning. Reduced nutrient uptake by plants can result in lesser crop yields and decreased productivity in agricultural systems. Furthermore, disruptions to mycorrhizal fungi populations can destabilize soil ecosystems, making them more susceptible to erosion, nutrient loss, and invasion by opportunistic species. This underscores the importance of considering the broader ecological consequences of pesticide use and adopting integrated pest management strategies that minimize harm to beneficial soil organisms like mycorrhizal fungi (Bhat et al. 2017). In addition to this, "agricultural practices such us tillage, crop rotation, fertilization, pesticide application, and irrigation can also reduce root colonization by mycorrhizal fungi." Thus, the use of agrochemicals needs to be addressed on priority basis, and as such, there is a dire need of replacement of agrochemicals with organic fertilizers.

Contrarily, the detrimental impacts of chemical fungicides on human health and the environment were highlighted by Haggag and Mohamed (2007). Prolonged use of chemical fungicides can result in the development of resistance in phytopathogenic fungi, rendering the fungicides ineffective, as noted by Kim and Hwang (2007). This resistance necessitates the use of alternative fungicides for effective disease control. In response to these challenges, the exploitation of microbes as biocontrol agents to manage phytopathology is proposed as a capable alternative scheme, as suggested by Kulkarni et al. (2007). Microorganisms represent diverse paragon of species, complex ecosystem interactions, and several metabolic conduits, making them valuable resources for biological activity, as emphasized by Tejesvi et al. (2007) and Raghukumar (2008). By harnessing the natural antagonistic properties of certain microorganisms, biological control agents can effectively suppress plant pathogens and reduce the need for chemical fungicides. This approach not only minimizes the negative impacts on human well-being and the environmental quality but also helps mitigate the advancement of fungicide resistance in phytopathogens. Overall, the exploitation of microorganisms for biological disease control represents a promising strategy for sustainable agriculture, leveraging the inherent capabilities of microbes to combat plant diseases while reducing reliance on synthetic chemicals.

Trichoderma species have indeed garnered significant attention as biocontrol agents due to their capacity to subdue soilborne plant pathogens through various mechanisms, including competition for nutrients and space, antibiosis, and induction of systemic resistance in plants. Their effectiveness in controlling diseases caused by fungi, bacteria, and nematodes has made them popular choices for integrated pest management strategies in agriculture. *Trichoderma* species like *T. harzianum*, *T. virens*, and *T. viride* have been extensively studied and commercialized for their biocontrol potential. It's remarkable to see how research in this field has advanced over the years, offering sustainable alternatives to conventional chemical pesticides (Benítez et al. 2004).

6.3   Organic Farming: A Way Forward to Good Health and Sustainable Development

"Organic farming" has achieved pedestal globally and has stretched in the recent decades due to ecological, fiscal, and societal concerns. "Organic farming" has been projected as a viable substitute in agricultural setup to facilitate the environmental tribulation's solution which basically stems out from conformist execution, for example, "frequent pesticide applications, excessive inputs of chemical fertilizers, soil degradation and the presence of pesticide residues in food" (Stockdale et al. 2001).

Organic farming has gained prominence as a holistic approach to agriculture that addresses a range of interconnected issues including environmental sustainability, public health, and economic viability (Ozturk et al. 2022). The unique characteristics of organic farming are as follows:

6.3.1   Ecological Concerns

With growing awareness of environmental degradation and the impacts of conventional agricultural practices on ecosystems, many people are turning to organic farming as a more sustainable alternative. Organic farming focuses on minimizing the use of synthetic inputs; therefore, it reduces environmental pollution. It gives impetus to soil health, biodiversity conservation, and natural resource management.

6.3.2   Economic Considerations

Organic farming offers economic benefits for both farmers and consumers. During transitioning to organic farming, it demands initial investment and adjustments and often leads to long-term cost savings by condensing the requirement for expensive synthetic inputs such as pesticides and fertilizers. Additionally, organic products typically are available on higher prices in the marketplace due to consumer demand for healthier and more environmentally friendly food options.

6.3.3   Societal Trends

There is an increased consumer demand for organic products driven by concerns about food safety, health, and sustainability. Consumers are increasingly choosing organic foods as a way to reduce exposure to synthetic chemicals and support environmentally friendly agricultural practices. This demand has stimulated growth in the organic food market, encouraging more farmers to transition to organic production methods.

6.3.4   Policy and Support

Government policies and programs promoting organic agriculture have also contributed to its expansion. Many countries offer incentives, subsidies, and certification programs to support organic farming and encourage the adoption of organic practices. These policies reflect a recognition of the environmental and societal benefits of organic farming and aim to incentivize its further development.

6.4   Organic Production Systems

Organic production systems have been escalating with an "annual growth rate of 26%" and presently assume a progressive job in agriculture in the "European Union." The evolution from conformist to "organic farming" often results in momentous transformations in the soil chemical attributes which probably amend soil fertility and mineral accessibility to crops either unswervingly by contributing to nutrient pools or circuitously by manipulating the soil environ. Discourses contrasting the "conformist and organic farming systems" have revealed amplification in "soil organic matter (SOM)" and mineral contents in organically managed soils. The practical applicability of organic farming for sustainable development is shown in Fig. 6.1.

Various advantageous strategies of organic production systems for food safety and environmental conservation include the following:

(i) The novel strategies have put emphasis on producing nutrient-rich, high-quality food in an eco-friendly manner reflecting a broader recognition of the interconnectedness between food production, environmental sustainability,

Fig. 6.1 An eco-friendly approach of organic farming

and public health (Raja 2013). By adopting these novel approaches, stakeholders in the food system aim to meet the growing demand for nutritious food while safeguarding the health of the planet and its inhabitants. The concept of rotational soil conditioning, as mentioned by Araujo et al. (2008), underscores the importance of using natural inputs to enhance soil fertility and manage fields effectively. This approach aligns with the broader trend toward sustainable agriculture, where minimizing reliance on synthetic chemicals and maximizing the use of natural resources are key priorities.

(ii) Organic farming practices, as outlined by Megali et al. (2013), advocate the use of natural inputs and techniques that promote soil health and ecosystem balance. By avoiding synthetic pesticides and fertilizers, organic farming helps preserve soil fertility and biodiversity, contributing to the long-term sustainability of agricultural systems. Additionally, the advantage of biofertilizers is that these are derived from organic materials such as compost, manure, or microbial inoculants. Biofertilizers, as noted by Sahoo et al. (2014), offer supplementary benefits such as longer shelf life and minimal adverse effects on ecosystems compared to synthetic fertilizers. These biofertilizers provide essential nutrients to plants while promoting beneficial microbial activity in the soil, further enhancing soil fertility and plant health in an environmentally friendly manner.

(iii) Organic farming often relies on harnessing the natural soil biodiversity to support plant growth and suppress diseases. Plant growth-promoting rhizobacteria (PGPR) and arbuscular mycorrhizal fungi (AMF) are key components of this soil microflora. Despite of their petite volume in soil, microorganisms are key players in the cycling of nitrogen, sulfur, and phosphorus and decomposition of organic residues and thus are also called hidden beauties. Heavy metals at elevated concentration are known to effect soil microbial population and their associated activities, which may directly influence the soil fertility.

(iv) The soil microbial population faces immense challenges due to contamination by various toxic substances. Xenobiotics from the sewage sludge and wastewater can exert significant stress on soil microorganisms, disrupting their populations and functions (Haq et al. 2021). Injudicious use of chemicals has resulted in multidimensional environmental problems in addition to descent of the food value of the agricultural produce. As such, there is a dire need of using organic formulations that have lesser impacts on our agroecosystems. Biofertilizers play a crucial role in maintaining soil fertility and promoting sustainable agriculture by harnessing the beneficial activities of microorganisms. These formulations contain living microorganisms that enhance nutrient availability, promote plant growth, and suppress soilborne diseases through various mechanisms (Solomon et al. 2023).

(v) Singh et al. (2011) represented valuable comprehension regarding the role of biofertilizers in enhancing crop productivity through improved nutrient cycling. When biofertilizers are applied as seed or soil inoculants, they establish and proliferate in the rhizosphere, the soil region influenced by plant roots. In this environment, biofertilizer microorganisms interact with the plant roots and soil constituents, contributing to various processes that promote nutrient availability and uptake, ultimately leading to improved crop productivity. Unlike chemical fertilizers, which often result in significant nutrient loss, biofertilizers offer a more sustainable approach to nutrient management by facilitating efficient nutrient uptake by plants. Furthermore, Adesemoye and Kloepper (2009) emphasize that microbial inoculants play a crucial role in sustaining agricultural productivity and fostering environmental health. By promoting nutrient cycling, improving soil fertility, and enhancing plant growth, microbial inoculants contribute to the overall sustainability of agricultural systems.

(vi) Mendes et al. (2013) highlighted the remarkable microbial diversity present in the rhizosphere, the soil region directly influenced by plant roots. Their findings indicate that the rhizosphere can host an extraordinarily large number of microbial cells, reaching densities of up to 10^{11} cells per gram of root. Moreover, this microbial community is incredibly diverse, potentially encompassing over 30,000 species of prokaryotes. These microorganisms play a crucial role in enhancing plant productivity through various mechanisms. The collective genome of the microbial community in the rhizosphere, known as the microbiome, is larger than that of plants themselves, as noted by Bulgarelli et al. (2013). This microbiome interacts with plant roots and influences crop health in natural agroecosystems by providing a range of beneficial services. In this context, Berg et al. (2013) aptly emphasize the indispensable services of soil microorganisms in agricultural ecosystems. Their contributions are essential for maintaining soil health, enhancing plant growth, and sustaining overall ecosystem function.

6.5   Soil Microbial Biodiversity as a Chauffeur of Organic Farming

It is pertinent to highlight various agriculturally useful microbial populations that play important roles in promoting plant growth, enhancing soil fertility, and protecting plants from diseases and environmental stresses. According to Singh et al. (2011), these microbial populations include the following:

6.5.1   Plant Growth-Promoting Rhizobacteria (PGPR)

PGPR represents a group of beneficial bacteria that inhabit the soil root zone. PGPR improves the growth and health of plants by different mechanisms. By colonizing the rhizosphere and employing these beneficial mechanisms, PGPR contribute to improved plant growth, stress tolerance, and nutrient acquisition, ultimately enhancing crop productivity and sustainability. Incorporating PGPR-based biofertilizers or bioinoculants into agricultural practices can help optimize plant-microbe interactions and promote eco-friendly approaches to crop production.

6.5.2 N_2-Fixing Cyanobacteria

Cyanobacteria are photosynthetic microorganisms and known for biological nitrogen fixation. They contribute to soil fertility by enriching it with nitrogen.

6.5.3   Mycorrhiza

The symbiotic associations with plant roots, augmenting the nutrient uptake, particularly phosphorus, in exchange for carbohydrates from the plant. Mycorrhizae improve soil structure and contribute to plant health and productivity.

6.5.4   Plant Disease-Suppressive Beneficial Bacteria

Certain bacteria have the capability to suppress phytopathology by competing with or antagonizing pathogens, enhancing plant defense mechanisms, or producing antibiotics or other inhibitory compounds.

6.5.5   Stress Tolerance Endophytes

Endophytic microorganisms exist inside plant tissues without causing damage and aid plants to endure environmental stresses such as drought, salinity, or heat shock. They may enhance plant growth and yield under adverse conditions.

6.5.6   Biodegrading Microbes

Microorganisms capable of degrading organic compounds play a critical character in recycling nutrients and breaking down of organic matter.

Therefore, by providing these valuable services, agriculturally useful microbial populations contribute to sustainable agriculture by promoting soil health, enhancing crop productivity, reducing environmental impacts, and fostering resilience in agricultural systems. Incorporating microbial-based products and practices into agricultural management strategies can help optimize these beneficial interactions and support the transition toward more sustainable and environmentally friendly farming practices. Thus, understanding and harnessing the potential of these microbes can lead to the progression of more effective and sustainable agricultural practices.

The integration of biofertilizers with traditional soil and crop management practices offers a promising approach to augment soil fertility, promote plant growth, and increase crop production in a sustainable manner. Biofertilizers, which consist of beneficial microorganisms like nitrogen-fixing bacteria, phosphate-solubilizing bacteria, and mycorrhizal fungi, offer several advantages when integrated with other sustainable farming techniques. Likewise, several types of PGPR have been documented to enhance soil attributes and crop productivity, particularly under no tillage or minimum tillage treatments. Aziz et al. (2012) documented the diversity of agriculturally beneficial microorganisms commonly used as biofertilizers. The findings of Dhanasekar and Dhandapani (2012) highlights the beneficial effects of competent strains of *Rhizobacter*, *Azotobacter*, *Phosphobacter*, and *Azospirillum* on growth and yield of sunflower (*Helianthus annuus*). Correspondingly, in rice cultivation, the incorporation of *Azotobacter*, *Azospirillum*, and *Rhizobium* has been found to promote plant physiology and improve root morphology.

Azotobacter plays a crucial role in the biological nitrogen fixation, contributing to various metabolic functions essential for soil fertility and plant growth, as highlighted by Sahoo et al. (2013). In addition to nitrogen fixation, Revillas et al. (2000) noted that *Azotobacter* has the ability to produce important vitamins such as thiamine (vitamin B1) and riboflavin (vitamin B2). Additionally, *Azotobacter* can produce phytohormones including indole acetic acid, gibberellins, and cytokinins, as demonstrated by AbdEl-Fattah et al. (2013). Gholami et al. (2009) observed that *Azotobacter*, especially *A. chroococcum*, exerts multiple beneficial effects on plant growth and health by promoting seed germination, improving root architecture, and suppressing the growth of pathogenic microorganisms, *Azotobacter* enhances plant growth and health, ultimately contributing to increased crop productivity. In nutshell, *Azotobacter* biofertilizers offer multiple benefits for agricultural systems, including nitrogen fixation, vitamin and hormone production, promotion of plant growth, and suppression of soilborne pathogens, making them valuable components of sustainable crop management practices.

Azospirillum is indeed a significant bacterium in agricultural contexts, as reported by Sahoo et al. (2014). It's known for its adaptability to flooded conditions, making it particularly valuable in various agricultural settings. Scientific research by Bhattacharyya and Jha (2012) indicates that *Azospirillum* promotes numerous characteristics of plant growth and development. Saikia et al. (2013) provide further evidence supporting the positive effects of *Azotobacter* on crop yields through greenhouse and field trials. These findings corroborate the observations of Gholami et al. (2009) regarding the beneficial impact of *Azotobacter*, particularly *A. chroococcum*, on plant growth and productivity. Notably, *Azospirillum* inoculation can modify root morphology by producing PGPR substances, as observed by Bashan et al. (2004). One mechanism through which *Azospirillum* enhances plant growth is through siderophore production. Mehdipour-Moghaddam et al. (2012) provide valuable insights into the mechanisms by which *Azotobacter* enhances plant growth by increasing the number of lateral roots and enhancing root hair formation. These adaptations result in a greater root surface area available for nutrient absorption, ultimately contributing to improved plant growth and nutrient uptake. Improvements in water status and nutrient profiles resulting from *Azospirillum* inoculation further advance plant growth and development, as reported by Ilyas et al. (2012). Overall, *Azospirillum*'s ability to promote plant growth, modify root morphology, and enhance nutrient uptake makes it a valuable tool in sustainable agricultural practices aimed at improving crop productivity and resilience (Dar et al. 2020).

Rhizobium is indeed a valuable nitrogen-fixing bacterium with significant contributions to agricultural productivity, as highlighted by Sharma et al. (2011). Its capability to transform atmospheric nitrogen into usable forms is crucial for enhancing crop yields. *Rhizobium*'s resilience to different temperature ranges allows it to effectively colonize plant roots, multiply within root hairs, and form nodules, as noted by Nehra et al. (2007). This symbiotic relationship between *Rhizobium* and legume plants enables the efficient utilization of atmospheric nitrogen by the plant. Several studies have validated the positive impact of *Rhizobium* inoculation on grain yields across various crops and soil types. For example, Patil and Medhane (1974) reported significant increases in grain yields of Bengal gram, while Rashid et al. (2012) observed similar effects in lentils. *Rhizobium* inoculants have also been shown to enhance yields in crops such as pea, alfalfa, sugar beet, berseem, groundnut, and soybean, as documented by Sharma et al. (2011) and Grossman et al. (2011). Moreover, Peng et al. (2008) reported that *Rhizobium* isolates acquired from wild rice exhibit the capability to supply nitrogen to cultivated rice plants, thereby promoting their growth and development. This finding highlights the potential of wild rice-associated *Rhizobium* strains to enhance nitrogen availability and improve the performance of cultivated rice crops. In organic farming, the ability of *Rhizobium* to enhance crop yields makes it a key player in sustainable agricultural practices aimed at improving soil fertility and reducing reliance on synthetic fertilizers.

Switching over to organic farming will eventually lead to sustainable agriculture and environmental conservation. It is imperative to encourage organic farming to counteract the detrimental consequences of agricultural xenobiotics for sustainable progression.

References

AbdEl-Fattah DA, Ewedab WE, Zayed MS, Hassaneina MK (2013) Effect of carrier materials, sterilization method, and storage temperature on survival and biological activities of *Azotobacter chroococcum* inoculants. Ann Agric Sci 58:111–118

Adesemoye AO, Kloepper JW (2009) Plant-microbes interactions in enhanced fertilizer-use efficiency. Appl Microbiol Biotechnol 85:1–12

Araujo ASF, Santos VB, Monteiro RTR (2008) Responses of soil microbial biomass and activity for practices of organic and conventional farming systems in Piaui state, Brazil. Eur J Soil Biol 44:225–230

Aziz G, Bajsa N, Haghjou T, Taule C, Valverde A, Mariano J, Arias A (2012) Abundance, diversity and prospecting of culturable phosphate solubilizing bacteria on soils under crop–pasture rotations in a no-tillage regime in Uruguay. Appl Soil Ecol 61:320–326

Bashan Y, Holguin G, Bashan LE (2004) *Azospirillum*-plant relationships: agricultural, physiological, molecular and environmental advances (1997–2003). Can J Microbiol 50:521–577

Benítez T, Rincón MA, Limón MC, Codón CA (2004) Biocontrol mechanisms of *Trichoderma* strains. Int Microbiol 7:249–260

Berg G, Zachow C, Müller H, Phillips J, Tilcher R (2013) Next-generation bio-products sowing the seeds of success for sustainable agriculture. Agronomy 3:648–656

Bhat RA, Dervash MA, Mehmood MA, Bhat MS, Rashid A, Bhat JIA, Singh DV, Lone R (2017) Mycorrhizae: a sustainable industry for plant and soil environment. In: Varma A et al (eds) Mycorrhiza—nutrient uptake, biocontrol, ecorestoration. Springer International Publishing, pp 473–502

Bhattacharyya PN, Jha DK (2012) Plant growth-promoting rhizobacteria (PGPR): emergence in agriculture. World J Microbiol Biotechnol 28:1327–1350

Bulgarelli D, Schlaeppi K, Spaepen S, Loren V, van Themaat E, Schulze-Lefert P (2013) Structure and functions of the bacterial microbiota of plants. Annu Rev Plant Biol 64:807–838

Dar SA, Bhat RA, Dervash MA (2020) Azotobacter as biofertilizer for sustainable soil and plant health under saline environmental conditions. In: Microbiota and biofertilizers: a sustainable continuum for plant and soil health. Springer, pp 231–254

Dhanasekar R, Dhandapani R (2012) Effect of biofertilizers on the growth of *Helianthus annus*. Int J Plant Ani Environ Sci 2:143–147

Gholami A, Shahsavani S, Nezarat S (2009) The effect of plant growth promoting rhizobacteria (pgpr) on germination seedling growth and yield of maize. Int J Biol Life Sci 5:1

Grossman JM, Schipanski ME, Sooksanguan T, Drinkwater LE (2011) Diversity of rhizobia nodulating soybean Glycine max (Vinton) varies under organic and conventional management. Appl Soil Ecol 50:14–20

Haggag WM, Mohamed HAA (2007) Biotechnological aspects of microorganisms used in plant biological control. Am-Eurasian J Sustain Agric 1:7–12

Haq S, Bharose R, Bhat RA, Ozturk M, AltayV BAA, Dervash MA, Hakeem KR (2021) Impact of treated sewage water on nutrient status of Alfisols and vegetable crops. Notulae Botanicae Horti Agrobotanici Cluj-Nappoca 49:12255

Ilyas N, Bano A, Iqbal S, Raja NI (2012) Physiological, biochemical and molecular characterization of *Azospirillum* spp. isolated from maize under water stress. Pak J Bot 44:71–80

Kim BS, Hwang BK (2007) Microbial fungicides in the control of plant diseases. J Phytopathol 155:641–653

Kulkarni M, Chaudhari R, Chaudhari A (2007) Novel tensio-active microbial compounds for biocontrol applications. In: General concepts in integrated pest and disease management (eds. A. Ciancio and K.G. Mukerji). Springer. Pp. 295–304

Liebich J, Schäffer A, Burauel P (2003) Structural and functional approach to studying pesticide side-effects on specific soil functions. Environ Toxicol Chem 22:784–790

Mahankale NR (2024) Global influence of synthetic fertilizers on climate change. Appl Geomat 16:317. https://doi.org/10.1007/s12518-023-00511-0

Megali L, Glauser G, Rasmann S (2013) Fertilization with beneficial micro-organisms decreases tomato defenses against insect pests. Agron Sustain Dev. https://doi.org/10.1007/s13593-013-0187-0

Mehdipour-Moghaddam MJ, Emtiazi G, Salehi Z (2012) Enhanced auxin production by *Azospirillum* pure cultures from plant root exudates. J Agric Sci Technol 14:985–994

Mendes R, Garbeva P, Raaijmakers JM (2013) The rhizosphere microbiome: significance of plant beneficial plant pathogenic and human pathogenic microorganisms. FEMS Microbiol Rev 37:634–663

Menendez A, Martínez A, Chiocchio V, Venedikian N, Ocampo JA, Godeas A (1999) Influence of the insecticide dimethoate on arbuscular mycorrhizal colonization and growth in soybean plants. Int Microbiol 2:43–45

Nehra K, Yadav SA, Sehrawat AR, Vashishat RK (2007) Characterization of heat resistant mutant strains of *rhizobium* sp. [Cajanus] for growth, survival and symbiotic properties. Indian J Microbiol 47:329–335

Ozturk M, Akram NA, Turkyilmaz BU, Ashraf M (2022) Introduction and application of organic fertilizers as protectors of our environment. Cambridge Scholars Publishing

Pal R, Chakrabarti K, Chakraborty A, Chowdhury A (2005) Pencycuron application to soils: degradation and effect on microbiological parameters. Chemosphere 60:1513–1522

Patil PL, Medhane NS (1974) Seed inoculation studies in gram (*Cicer arietinum*) with different strains of *rhizobium* sp. Plant Soil 40:221–223

Peng G, Yuan Q, Li H, Zhang W, Tan Z (2008) *Rhizobium oryzae* sp. nov., isolated from the wild rice *Oryza alta*. Int J Syst Evol Microbiol 58:2158–2163

Potera C (2007) Agriculture: pesticides disrupt nitrogen fixation. Environ Health Perspect 115(12):A579

Raghukumar C (2008) Marine fungal biotechnology: an ecological perspective. Fungal Divers 31:19–35

Raja N (2013) Biopesticides and biofertilizers: ecofriendly sources for sustainable agriculture. J Biofertil Biopestici 4:13. https://doi.org/10.4172/2155-6202.1000e112

Rashid MH, Schafer H, Gonzalez J, Wink M (2012) Genetic diversity of rhizobia nodulating lentil (*Lens culinaris*) in Bangladesh. Syst Appl Microbiol 35:98–109

Revillas JJ, Rodelas B, Pozo C, Martinez-Toledo MV, Gonzalez LJ (2000) Production of B-group vitamins by two *Azotobacter* strains with phenolic compounds as sole carbon source under diazotrophic and adiazotrophic conditions. J Appl Microbiol 89:486–493

Sahoo RK, Ansari MW, Dangar T, Mohanty S, Tuteja N (2013) Phenotypic and molecular characterization of efficient nitrogen fixing *Azotobacter* strains of the rice fields. Protoplasma 251:511–523. https://doi.org/10.1007/s00709-013-0547-2

Sahoo RK, Ansari MW, Pradhan M, Dangar TK, Mohanty S, Tuteja N (2014) Phenotypic and molecular characterization of efficient native *Azospirillum* strains from rice fields for crop improvement. Protoplasma 251:943–p53. https://doi.org/10.1007/s00709-013-0607-7

Saikia SP, Bora D, Goswami A, Mudoi KD, Gogoi A (2013) A review on the role of *Azospirillum* in the yield improvement of non-leguminous crops. Afr. J Microbiol Res 6:1085–1102

Sharma P, Sardana V, Kandola SS (2011) Response of groundnut (*Arachis hypogaea* L.) to rhizobium inoculation. Libyan Agric Res Centre J Int 2:101–104

Singh JS, Pandey VC, Singh DP (2011) Efficient soil microorganisms: a new dimension for sustainable agriculture and environmental development. Agric Ecosyst Environ 140:339–353

Solomon W, Mutum L, Janda T et al. (2023) Potential benefit of microalgae and their interaction with bacteria to sustainable crop production. Plant Growth Regul 101:53–65. https://doi.org/10.1007/s10725-023-01019-8

Stamenković S, Beškoski V, Karabegović I, Lazić M, Nikolić N (2018) Microbial fertilizers: a comprehensive review of current findings and future perspectives. Span J Agric Res 16:e09R01

Stockdale EA, Lampkin NH, Hovi M, Keatinge R, Lennartsson EKM, Macdonald DW, Padel S, Tattersall FH, Wolfe MS, Watson CA (2001) Agronomic and environmental implications of organic farming systems. Adv Agron 70:261–327

Tejesvi MV, Kini KR, Prakash HS, Subbiah V, Shetty HS (2007) Genetic diversity and antifungal activity of species of *Pestalotiopsis* isolated as endophytes from medicinal plants. Fungal Divers 24:37–54

Wei X, Xie B, Wan C, Song R, Zhong W, Xin S, Song K (2024) Enhancing soil health and plant growth through microbial fertilizers: mechanisms, benefits, and sustainable agricultural practices. Agronomy 14:609. https://doi.org/10.3390/agronomy14030609

Chapter 7
Fungal Internet: *The Natural Networking Systems*

Abstract The term "fungal internet," coined by Dr. Suzanne Simard and popularized by Dr. Paul Stamets, describes the intricate networks of mycorrhizal fungi connecting trees and facilitating communication and resource exchange in forest ecosystems. These "wood wide web" networks enable nutrient transfer, chemical signaling, and disease resistance among trees, illustrating the profound interconnectedness and complexity of forest ecosystems. Highlighting the sophisticated interactions between fungi and plants, this concept underscores the vital role of fungal networks in maintaining ecosystem health and resilience. Furthermore, in this chapter, the comprehension has been made more pragmatic through case studies on fungal internet. Thus, exploration and research on the functions of fungal internet is crucial for ecological management, conservation strategies, and enhancing pollution remediation efforts by leveraging fungi's ability to improve soil health and plant resilience.

7.1 Introduction

The term "fungal internet" was coined by Dr. Suzanne Simard, a professor of forest ecology at the University of British Columbia, Canada (Simard et al. 1997). She used this term to describe the complex network of mycorrhizal fungi that connect trees and facilitate communication and resource exchange in forest ecosystems. Dr. Simard's pioneering research has shed light on the importance of these fungal networks, often referred to as the "wood wide web," in facilitating nutrient transfer, chemical signaling, and even disease resistance among trees. Her work has significantly influenced our understanding of the interconnectedness of forest ecosystems and the role of fungi in maintaining their health and resilience. However, the idea of the "fungal internet" was popularized by a comprehensive study of Dr. Paul Stamets on the *Mycelium Running* (Stamets 2005). By this term, he means the complex system of mycelium, threadlike structures that compose the fungi and their connections across the plants. The fungi help the plants share the nutrients and chemical signals

© The Author(s), under exclusive license to Springer Nature Switzerland AG 2024 77
M. A. Dervash et al., *Soil Organisms*, SpringerBriefs in Microbiology,
https://doi.org/10.1007/978-3-031-66293-5_7

and eventually warn each other about threats such as pests and disease. This example illustrates the complexity of relationships in the ecosystem and the interconnectedness of all organisms.

Indeed, the "fungal internet" or "fungal network" is a fascinating example of natural communication and resource-sharing systems, showcasing the complexity and interconnectedness of life beneath our feet. This invisible network not only underscores the sophistication of fungal and plant interactions but also highlights the importance of preserving soil health and integrity in ecological management and conservation strategies. Mycorrhizal fungi play a significant role in ecosystem dynamics and have shown potential in aiding pollution remediation. This remediation process can be enhanced by understanding the role fungi play in soil health, plant resilience, and the degradation or stabilization of pollutants.

Furthermore, in this chapter, the comprehension has been made more pragmatic through case studies on fungal internet. Thus, exploration and research on the functions of fungal internet are crucial for ecological management, conservation strategies, and enhancing pollution remediation efforts by leveraging fungi's ability to improve soil health and plant resilience.

7.2   Understanding Fungal (Mycorrhizal) Internet

Mycorrhizal fungi form symbiotic relationships with the roots of most plant species. In this mutualistic association, the fungus colonizes the plant's root system and extends its hyphae far into the soil. These hyphae form an extensive underground network that can link many plants together (Begum et al. 2019).

Fungal networks consist of mycelium, the vegetative part of fungi, which extends through the soil in the form of threadlike structures known as hyphae. These networks can be extensive, covering large areas and connecting many plants together, including different species. The two main types of mycorrhizal associations are (Teste et al. 2020) as follows:

(i) *Arbuscular mycorrhizal fungi (AMF)*: These fungi penetrate the root cells of plants, forming structures called arbuscules, which are essential for the exchange of nutrients between the fungus and the plant. AMF are common in agricultural and grassland ecosystems. AMF primarily help in the uptake of phosphorus but also assist in the absorption of other micronutrients.

(ii) *Ectomycorrhizal fungi*: These fungi form a sheath around plant roots and do not penetrate the root cells directly. They are particularly important in forest ecosystems and are associated with trees like pines and oaks. Ectomycorrhizal fungi are crucial for the cycling of nitrogen and phosphorus.

7.3   Fungal Networks and Nutrient Cycling

Fungal networks, particularly those formed by mycorrhizal fungi, play a pivotal role in nutrient cycling within ecosystems. These networks not only facilitate the exchange of nutrients between the soil and plant roots but also enhance the movement of nutrients across the landscape, influencing the productivity and health of plant communities (Simard 2018).

7.4   Role in Nutrient Cycling

Fungal networks enhance nutrient cycling in several key ways:

(i) *Enhanced Nutrient Uptake*: Mycorrhizal fungi increase the surface area in contact with the soil, allowing plants to access a larger volume of soil and thus more nutrients than the roots alone could reach. This is particularly important for the uptake of phosphorus, which is relatively immobile in the soil.

(ii) *Nutrient Transport*: Nutrients can be transported through the fungal hyphae from areas of high concentration (e.g., decomposing matter) to areas of low concentration (e.g., nutrient-poor soil). This redistribution can support plant growth in less fertile areas and contribute to the overall productivity of the ecosystem.

(iii) *Decomposition and Organic Matter Breakdown*: Saprotrophic fungi, which are not mycorrhizal but are part of the broader fungal community, play a critical role in breaking down organic matter. This process releases nutrients locked in dead plants and animals, making them available again for use by living plants.

(iv) *Carbon Cycling*: By interacting with plant roots, mycorrhizal fungi also influence carbon cycling. They help stabilize carbon in the soil by incorporating it into their biomass and the soil organic matter. This can affect soil carbon storage and the overall carbon dynamics of the ecosystem.

(v) *Linking Aboveground and Belowground Ecosystems*: Fungal networks mediate interactions not only among plants (e.g., sharing nutrients and signaling) but also between plants and other soil organisms, including bacteria, protozoa, and invertebrates. This contributes to a complex web of interactions that drive ecosystem functioning.

7.5   Ecological and Environmental Impact

The activities of fungal networks are crucial for maintaining the stability and resilience of ecosystems. They help sustain plant health and productivity, especially under stress conditions such as drought or nutrient-poor soils (Wahab et al. 2023).

Moreover, understanding these networks can aid in ecosystem management and conservation, helping to restore degraded lands and maintain soil fertility.

In agriculture, enhancing mycorrhizal associations through practices like reduced tillage, cover cropping, and the use of fungal inoculants can improve crop yield, reduce the need for chemical fertilizers, and increase soil health and sustainability (Wahab et al. 2023). Therefore, fungal networks are fundamental components of the biosphere's nutrient cycling processes. Their study not only sheds light on the intricate relationships within ecosystems but also opens avenues for ecological management and conservation efforts that support global food security and environmental sustainability.

7.6   Chemical Signaling Through Fungal Internet

Chemical signaling plays a vital role in the functioning of the fungal internet, allowing trees to communicate with each other through the exchange of signaling molecules via mycorrhizal networks.

7.6.1   Case Study of Douglas Fir (**Pseudotsuga menziesii***) and Paper Birch (**Betula papyrifera***) Interaction*

Research by a scientific team has demonstrated how Douglas fir and paper birch trees utilize mycorrhizal networks to exchange resources and information (Leanne et al. 2010). In particular, they found that Douglas fir seedlings, when shaded, receive carbon from neighboring paper birch trees through shared mycorrhizal fungi. Chemical signaling is crucial in this interaction. When the paper birch trees are exposed to sunlight, they photosynthesize and produce excess sugars. Some of these sugars are transferred to the roots and released into the soil. These sugars act as signaling molecules that attract mycorrhizal fungi. The mycorrhizal fungi colonize the roots of both paper birch and Douglas fir trees, forming a network that connects them underground. Meanwhile, Douglas fir seedlings in the shaded understory release chemical signals indicating their need for carbon. The mycorrhizal fungi pick up these signals and transfer the sugars obtained from paper birch to the Douglas fir seedlings. This exchange of resources through the fungal internet benefits both tree species. The paper birch trees indirectly receive nutrients from the Douglas fir seedlings in the form of nitrogen and other minerals obtained from deeper soil layers. This example illustrates how chemical signaling, facilitated by mycorrhizal networks, enables trees to communicate their needs and share resources across species boundaries. Such interactions enhance the resilience and productivity of forest ecosystems, highlighting the importance of the fungal internet in maintaining ecological balance and biodiversity (Leanne et al. 2010).

7.7 Disease Resistance by Fungal Internet

Disease resistance among trees facilitated by mycorrhizal networks, or the fungal internet, is an intriguing aspect of forest ecology. With the help of few case studies discussed as under, let us understand how fungal internet works in inducing disease resistance among the species:

7.7.1 Case Studies Pertaining to Disease Resistance by Fungal Internet

(i) *Western Red Cedar (Thuja plicata) and Douglas Fir (Pseudotsuga menziesii) Interaction*: A research has shown that Western red cedar can transfer defense signals through mycorrhizal networks to neighboring Douglas fir trees, enhancing their resistance to pathogens (Aghai et al. 2019). When Western red cedar trees are attacked by pathogens such as root rot fungi (e.g., *Armillaria* species), they produce defense compounds and signaling molecules. These defense signals are transmitted through the mycorrhizal fungi connecting the roots of Western red cedar and Douglas fir trees. Upon receiving these signals, Douglas fir trees activate their own defense mechanisms, which may include producing antifungal compounds or strengthening their cell walls to resist fungal invasion. As a result, Douglas fir trees connected to Western red cedar via mycorrhizal networks exhibit increased resistance to root rot pathogens. This reciprocal exchange of defense signals between tree species demonstrates how the fungal internet can enhance disease resistance in forest ecosystems (Aghai et al. 2019).

(ii) *Black Cottonwood (Populus trichocarpa) and Willow (Salix spp.) Interaction*: Black Cottonwood and Willow trees have been observed to share defense signals through mycorrhizal networks, improving their resistance to herbivores (Doty et al. 2009). When one of these tree species is attacked by herbivores such as leaf-chewing insects, it releases volatile organic compounds (VOCs) as part of its defense response. These VOCs can travel through the air or be absorbed by the soil and taken up by mycorrhizal fungi associated with neighboring trees. The mycorrhizal fungi then transmit these signals to the roots of other trees, including those of different species, such as black cottonwood and willow. In response to the received signals, the recipient trees may activate their own defense mechanisms, such as producing toxic compounds or increasing leaf toughness, to deter herbivores. By sharing defense signals through the fungal internet, black cottonwood and willow trees can collectively defend against herbivore attacks, thereby enhancing their survival and fitness.

These examples illustrate how mycorrhizal networks facilitate the exchange of defense signals among tree species, contributing to enhanced disease resistance and overall resilience in forest ecosystems. The interconnectedness enabled by the fungal internet plays a crucial role in maintaining the health and integrity of diverse forest communities.

7.8   Allied Functions of the Fungal Network

7.8.1   Resource Sharing

The fungal network can transport water, nitrogen, phosphorus, and other nutrients from areas of high availability to areas of low availability. This helps plants in nutrient-poor conditions to receive sustenance from better-situated individuals.

7.8.2   Communication

Plants use the mycorrhizal network to send and receive chemical signals. For instance, if one plant is attacked by pests, it can release chemical signals through the network, warning neighboring plants to activate their own defense mechanisms.

7.8.3   Stress Alleviation

During times of environmental stress, such as drought, the network can help distribute resources from plants in less affected areas to those in more affected areas, potentially improving the survival and health of the entire community.

7.8.3.1   Carbon Storage

By linking many plants together, these networks can influence the plant community's carbon dynamics, potentially impacting carbon storage and release in ecosystems.

7.9   Implications and Research

The discovery and study of these fungal networks have profound implications for ecology, forestry, agriculture, and our understanding of plant communication and community dynamics (Simard 2018). For instance, managing forests and agricultural systems with an awareness of mycorrhizal networks could lead to practices that enhance crop productivity, forest health, and biodiversity.

Researchers continue to explore the complexities of these networks, including how environmental changes (like climate change, pollution, or land-use change) affect their functionality and resilience. Understanding these networks is also crucial for conservation efforts, as disrupting the soil and fungal networks can have cascading effects on plant communities and entire ecosystems.

References

Aghai MM, Khan Z, Joseph MR, Stoda AM, Sher AW, Ettl GJ, Doty SL (2019) The effect of microbial endophyte consortia on *Pseudotsuga menziesii* and *Thuja plicata* survival, growth, and physiology across edaphic gradients. Front Microbiol 10:1353. https://doi.org/10.3389/fmicb.2019.01353

Begum N, Qin C, Ahanger MA, Raza S, Khan MI, Ashraf M, Ahmed N, Zhang L (2019) Role of arbuscular mycorrhizal fungi in plant growth regulation: implications in abiotic stress tolerance. Front Plant Sci 10:1068. https://doi.org/10.3389/fpls.2019.01068

Doty SL, Oakley B, Xin G, Jun Won K, Glenda S, Zareen N, Azra V, James S (2009) Diazotrophic endophytes of native black cottonwood and willow. Symbiosis 47:23–33. https://doi.org/10.1007/BF03179967

Leanne P, Simard S, Jones M (2010) Pathways for belowground C transfer between paper birch and Douglas-fir seedlings. Plant Ecol Divers 3:221–233. https://doi.org/10.1080/17550874.2010.502564

Simard SW (2018) Mycorrhizal networks facilitate tree communication, learning, and memory. In: Baluska F, Gagliano M, Witzany G (eds) Memory and learning in plants signaling and communication in plants. Springer, Cham. https://doi.org/10.1007/978-3-319-75596-0_10

Simard SW, Perry DA, Jones MD, Myrold DD, Durall DM, Molina R (1997) Net transfer of carbon between ectomycorrhizal tree species in the field. Nature 388:579–582. https://doi.org/10.1038/41557

Stamets P (2005) Mycelium running: how mushrooms can help save the world. Ten Speed Press, New York

Teste FP, Jones MD, Dickie IA (2020) Dual-mycorrhizal plants: their ecology and relevance. New Phytol 225:1835–1851. https://doi.org/10.1111/nph.16190

Wahab A, Muhammad M, Munir A, Abdi G, Zaman W, Ayaz A, Khizar C, Reddy SPP (2023) Role of arbuscular mycorrhizal fungi in regulating growth, enhancing productivity, and potentially influencing ecosystems under abiotic and biotic stresses. *Plants* 12:3102. https://doi.org/10.3390/plants12173102

References

Chapter 8
Biodynamic Agriculture: *Unknotting the Secrets of Sustainability*

Abstract Biodynamic agriculture is a holistic approach that views the farm as a self-sustaining ecosystem, emphasizing the interrelationships between soil, plants, animals, and cosmic influences. This agricultural system promotes environmental sustainability, high biodiversity, and natural resource conservation, while adhering to strict standards that exclude synthetic pesticides, fertilizers, GMOs, antibiotics, and growth hormones. Over recent decades, biodynamic agriculture has gained global traction due to growing environmental concerns and consumer preferences for organic products, integrating spiritual, ethical, and ecological principles into farming practices. In this chapter, global growth and trends with key principles, methods, merits, and demerits of biodynamic agriculture are discussed in detail. Besides, certification and challenges faced by biodynamic agriculture and the future vistas have also been discussed.

8.1 Introduction

The term "biodynamic agriculture" was coined by Rudolf Steiner in 1924. Steiner was an Austrian philosopher, social reformer, and esotericist who introduced this holistic approach to farming in a series of eight lectures given to farmers in 1924. These lectures were later compiled into a book titled *Spiritual Foundations for the Renewal of Agriculture* (Paull 2011). Biodynamic agriculture emphasizes a balanced relationship between soil, plants, animals, and cosmic influences, viewing the farm as a self-sustaining ecosystem. The global scenario of *biodynamic agriculture* has been evolving rapidly over the past few decades, with increasing awareness of environmental sustainability, health concerns, and consumer preferences shifting toward organic products.

Biodynamic agriculture is defined as a system of farm management and food production that combines best environmental practices, a high level of biodiversity, the preservation of natural resources, and the application of high animal welfare standards without the use of synthetic pesticides, fertilizers, genetically modified organisms (GMOs), antibiotics, and growth hormones. It incorporates spiritual, ethical, and ecological principles into agricultural practices. In this chapter, global growth and trends with key principles, methods, merits, and demerits of biodynamic agriculture are discussed in detail. Besides, certification and challenges faced by biodynamic agriculture and the future vistas have also been discussed.

8.2   Global Growth and Trends

8.2.1   *Expansion*

Biodynamic agriculture has seen substantial growth worldwide. According to reports from the Research Institute of Organic Agriculture, *Forschungsinstitut für biologischen Landbau* (FiBL), and *International Federation of Organic Agriculture Movements* (IFOAM)—Organics International, there has been a consistent increase in both the number of organic producers and the area of land under organic cultivation. As of the latest reports, over 180 countries practice biodynamic agriculture, with more than 71.5 million hectares managed organically (Willer et al. 2022).

8.2.2   *Leading Countries*

Australia (35.7×10^6 ha), Argentina (4.5×10^6 ha), and Uruguay (2.7×10^6 ha) are among the countries with the largest areas of organically managed agricultural land. Europe and North America also show significant organic market growth, driven by strong consumer demand and supportive policies (Willer et al. 2022).

8.2.3   *Market Size*

The global market for organic food and drink is expanding rapidly, with the United States, Germany, France, and China being the largest markets. Consumer demand continues to rise, fueled by increasing awareness of health and environmental issues.

8.3 Drivers of Biodynamic Agriculture

8.3.1 Health Concerns

Consumers are increasingly aware of the health implications of pesticide residues and other chemicals in conventional agriculture. Organic food is often perceived as healthier due to the absence of synthetic chemicals.

8.3.2 Environmental Awareness

Biodynamic agriculture supports greater biodiversity and aims to enhance soil health through practices such as crop rotations, organic fertilizers, and low-intensity pasture-based livestock husbandry. It is also associated with lower carbon emissions compared to conventional farming.

8.3.3 Government Policies

Many governments offer incentives for biodynamic agriculture through subsidies, research, and support for certification processes. Regulations such as the European Union's common agricultural policy (CAP) promote organic practices by providing financial assistance for organic certification and conversion (Yang et al. 2024).

8.3.4 Economic Incentives

Despite often higher prices, the premium market for organic products provides economic incentives for farmers to switch to organic methods. These premiums can help offset the typically lower yields and higher labor costs associated with biodynamic agriculture (Yang et al. 2024).

8.4 Science Behind Biodynamic Agriculture

Biodynamic agriculture is an ecological and holistic approach to farming that emphasizes the interconnectedness of soil, plants, animals, and the cosmos (Muhie 2022). Biodynamic farming incorporates spiritual, ethical, and ecological principles into agricultural practices. It is considered one of the earliest forms of organic

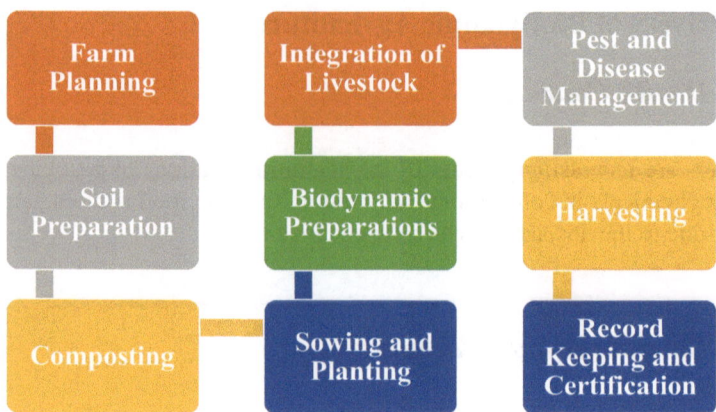

Fig. 8.1 Stepwise procedure for biodynamic farming

agriculture and focuses on enhancing soil fertility, biodiversity, and sustainability while producing high-quality, nutrient-dense food.

The schematic representation for stepwise procedure for biodynamic farming is represented in Fig. 8.1.

By following these steps, farmers can implement biodynamic practices that enhance ecological balance, soil fertility, and overall sustainability in their agricultural operations.

8.4.1   Key Principles of Biodynamic Agriculture

(i) *Holistic Perspective:* Biodynamic farming views the farm as a self-sustaining organism, with each component—soil, plants, animals, and humans—playing a vital role in the health and balance of the whole system.

(ii) *Biodiversity*: Maintaining biodiversity is central to biodynamic farming. Farms are encouraged to cultivate a variety of crops and incorporate livestock to enhance ecosystem resilience and nutrient cycling.

(iii) *Soil Health*: Biodynamic practices focus on building and maintaining healthy, living soils. This includes composting, cover cropping, crop rotation, and minimizing tillage to improve soil structure, fertility, and microbial activity.

(iv) *Cosmic Influences*: Biodynamic agriculture incorporates the influence of celestial rhythms and cosmic forces on farming practices. Planting, cultivation, and harvesting activities are timed according to lunar and planetary cycles.

(v) *Biodynamic Preparations*: Biodynamic farms use a series of specially prepared herbal and mineral-based preparations to enhance soil vitality and stimulate plant growth. These preparations are applied in homeopathic doses to compost piles, soil, and plants.

(vi) *Closed Nutrient Cycles*: Biodynamic farms aim to minimize external inputs and waste by recycling nutrients within the farm system. Composting, crop residues, and animal manures are used to replenish soil fertility and maintain nutrient cycles.

8.5 Practices in Biodynamic Agriculture

8.5.1 Compost Preparation

Biodynamic compost preparations, known as "BD preps," involve specific herbal and mineral mixtures that are added to compost piles to enhance microbial activity and nutrient availability.

8.5.2 Biodynamic Calendar

Planting, cultivating, and harvesting activities are guided by the biodynamic calendar, which takes into account lunar and planetary influences on plant growth and development.

8.5.3 Crop Rotation

Rotating crops helps maintain soil fertility, reduce pest and disease pressure, and prevent soil erosion. Biodynamic farms often implement diverse crop rotations tailored to their specific climate and soil conditions.

8.5.4 Livestock Integration

Integrating livestock, such as chickens, cows, and pigs, into the farm system helps recycle nutrients, improve soil fertility, and diversify farm income streams.

8.5.5 Seed Saving

Biodynamic farmers often save and exchange seeds to preserve traditional varieties and adapt them to local growing conditions, promoting seed diversity and resilience.

8.6   Different Approaches and Methods of Biodynamic Agriculture

Biodynamic agriculture encompasses various approaches and methods tailored to different climates, landscapes, and agricultural systems. Few examples of the types of biodynamic agriculture practiced around the world are as under:

8.6.1   Traditional Organic Agriculture

Traditional organic agriculture relies on crop rotation to maintain soil fertility and manage pests and diseases. Different crops are grown sequentially in the same field, with each crop providing different nutrients and disrupting pest and disease cycles. Organic farmers use compost, animal manure, and other organic materials to fertilize their crops. These materials provide nutrients for plant growth and improve soil structure and microbial activity. Traditional organic farmers rely on natural enemies of pests, such as predators, parasites, and beneficial insects, to control pest populations. This approach minimizes the use of synthetic pesticides and supports biodiversity on the farm.

8.6.2   Permaculture

Permaculture, a term coined by Bill Mollison and David Holmgren in the 1970s, is a design system that aims to create sustainable human habitats by modeling them after natural ecosystems. The term "permaculture" is a contraction of "permanent agriculture" or "permanent culture" (Clitheroe 2019). It integrates principles from ecology, agriculture, landscape design, and sustainable architecture to create systems that are self-sufficient, regenerative, and resilient:

(i) *Design Principles*: Permaculture is a holistic approach to farming that emphasizes working with nature to create sustainable and self-sufficient systems. Permaculture farms are designed based on principles such as observation, diversity, and integration of elements.

(ii) *Polyculture and Agroforestry*: Permaculture farms often use polyculture and agroforestry techniques to mimic natural ecosystems and maximize diversity. Different crops are grown together in a way that mimics natural forest ecosystems, enhancing biodiversity and soil health.

(iii) *Water Harvesting and Conservation*: Permaculture farms prioritize water harvesting and conservation techniques to maximize water efficiency and resilience to drought. Techniques such as swales, ponds, and mulching help capture and retain water in the landscape.

8.6.3 Community-Supported Agriculture (CSA)

The concept of community-supported agriculture (CSA) was developed by two Japanese women, *Jan Vander Tuin* and *Trauger Groh*, in the 1960s and 1970s. They were inspired by the *Teikei* movement in Japan, which translates to "food with the farmer's face on it" or "cooperative." This movement aimed to establish direct relationships between consumers and farmers, ensuring fair prices for farmers and access to fresh, locally grown produce for consumers (Jackson et al. 2011):

  (i) *Direct Sales to Consumers*: Community-supported agriculture (CSA) involves direct sales of produce to consumers through subscription-based memberships. Members receive regular deliveries of fresh produce from the farm, often with the option to visit the farm and participate in farm activities.

 (ii) *Shared Risk and Reward*: CSAs typically involve shared risk and reward between farmers and consumers. Members pay upfront for a season's worth of produce, providing farmers with financial stability, while farmers commit to providing high-quality, locally grown produce throughout the season.

(iii) *Local and Sustainable*: CSAs promote local and sustainable food systems by connecting consumers directly with local farmers. This reduces the carbon footprint associated with food transportation and supports small-scale, diversified farming operations.

Each approach has its own strengths and challenges, and farmers may combine elements from different approaches to suit their specific goals and conditions. Overall, biodynamic agriculture emphasizes sustainability, biodiversity, and environmental stewardship, offering a viable alternative to conventional agriculture.

8.7 Merits of Biodynamic Agriculture

Biodynamic agriculture plays a crucial role in benefiting soil environments in numerous ways, promoting soil health, fertility, and biodiversity while minimizing environmental impacts. Several key ways in which biodynamic agriculture contributes to the benefaction of soil environments include the following:

8.7.1 Soil Health and Fertility

  (i) *Enhanced Soil Structure*: Biodynamic agriculture practices such as reduced tillage, cover cropping, and organic matter additions improve soil structure, promoting better water infiltration, aeration, and root penetration.

 (ii) *Increased Soil Organic Matter*: Biodynamic agriculture relies on organic inputs such as compost, manure, and green manures, which add organic carbon to the soil. This organic matter serves as a source of energy for soil organisms and helps build soil fertility.

(iii) *Nutrient Cycling*: Biodynamic agriculture emphasizes closed-loop nutrient cycling by recycling organic materials within the farm system. Composting, crop rotations, and the use of leguminous cover crops contribute to nutrient availability and balance in the soil.

8.7.2   Soil Biodiversity

(i) *Microbial Communities*: Biodynamic agriculture practices support diverse and active microbial communities in the soil. These microbes play vital roles in nutrient cycling, disease suppression, and soil aggregation, contributing to overall soil health (Zeng et al. 2024).

(ii) *Macrofauna Diversity*: Organic farms often have higher levels of soil macrofauna such as earthworms, beetles, and ants. These organisms aid in soil aeration, nutrient cycling, and decomposition, enhancing soil structure and fertility (Dervash et al. 2018).

(iii) *Minimized Chemical Inputs*: Biodynamic agriculture avoids the use of synthetic pesticides, herbicides, and fertilizers, which can harm soil organisms and disrupt soil ecosystems. Instead, organic farmers rely on biological pest control, crop rotation, and other natural methods to manage pests and weeds (Muhie 2022).

(iv) *Water Quality*: By minimizing runoff of synthetic chemicals, biodynamic agriculture helps protect water quality and aquatic ecosystems. Healthy soils with good structure and organic matter content also enhance water retention and reduce erosion.

8.7.3   Reduced Environmental Impact

8.7.3.1   Climate Change Mitigation

(a) Biodynamic agriculture practices contribute to carbon sequestration in soils, helping mitigate climate change. Increased soil organic matter and reduced tillage can enhance carbon storage in soils, acting as a sink for atmospheric carbon dioxide (Santoni et al. 2022).

8.7.3.2   Resilience to Climate Change

(a) *Drought Resistance*: Healthy organic soils with improved structure and organic matter content are more resilient to drought and extreme weather events. They have better water-holding capacity and are less prone to erosion.
(b) *Adaptation*: Biodynamic agriculture promotes diverse cropping systems and genetic diversity, which can help farmers adapt to changing climatic conditions. Crop rotations, cover cropping, and agroforestry are examples of practices that enhance resilience.

8.7.3.3   Long-Term Sustainability

(a) *Soil Conservation*: Biodynamic agriculture practices prioritize soil conservation and long-term sustainability. By maintaining soil health and fertility, organic farms can continue to produce food while minimizing the need for external inputs.
(b) *Biodiversity Conservation*: Biodynamic agriculture supports biodiversity conservation by providing habitat and resources for diverse plant and animal species. This promotes ecological resilience and ecosystem services that benefit both agriculture and the environment.

8.8   Drawbacks of Biodynamic Agriculture

While biodynamic agriculture offers numerous benefits, it also has several drawbacks and challenges that farmers may encounter (Rigolot and Quantin 2022):

8.8.1   Lower Yields

Biodynamic agriculture often yields lower quantities of crops compared to conventional farming. This is primarily due to limitations on synthetic inputs, which can result in decreased pest and disease control and lower nutrient availability, leading to reduced productivity.

8.8.2   Higher Labor Costs

Biodynamic agriculture typically requires more labor-intensive practices such as hand weeding, crop rotation, and pest management. Without the use of synthetic pesticides and herbicides, farmers may need to invest more time and effort in managing pests and weeds manually, increasing labor costs.

8.8.3   Risk of Pest and Disease Damage

Biodynamic agriculture relies on natural and biological methods for pest and disease control, which may be less effective than synthetic chemical pesticides. As a result, organic crops may be more susceptible to pest and disease damage, leading to potential yield losses.

8.8.4   Limited Availability of Organic Inputs

Organic farmers face challenges in sourcing organic fertilizers, pesticides, and other inputs, especially in regions with limited access to organic products. This can result in higher costs and logistical difficulties in obtaining necessary inputs for biodynamic agriculture practices.

8.8.5   Transition Period

Converting from conventional to biodynamic agriculture requires a transition period during which farmers must adhere to organic standards but may not yet receive organic certification. This transition period can be financially challenging for farmers as they invest in organic practices without the premium prices associated with certified organic products.

8.8.6   Market Challenges

While demand for organic products is increasing, organic farmers may still face challenges in accessing markets and securing premium prices for their products. Market fluctuations and competition from conventional products can impact the profitability of biodynamic agriculture operations.

8.8.7   Certification Costs and Requirements

Obtaining organic certification can be costly and time-consuming, especially for small-scale farmers. Organic certification requires compliance with strict standards and documentation requirements, which may pose challenges for farmers, particularly those with limited resources or access to technical assistance.

8.8.8   Limited Crop Choices

Some crops may be more challenging to grow organically due to their susceptibility to pests and diseases or their nutrient requirements. Organic farmers may need to prioritize certain crops over others based on their suitability for organic production practices.

Though biodynamic agriculture offers numerous environmental and health benefits, it also presents challenges related to productivity, labor, input availability, and market access. Addressing these drawbacks requires ongoing research, innovation, and support for organic farmers to improve the viability and sustainability of organic agriculture.

8.9   Certification and Challenges Facing Biodynamic Agriculture

8.9.1   Yield Gap

Organic yields are generally lower than conventional yields, which is a significant challenge as global food demand increases. Research and innovation in biodynamic agriculture techniques are crucial for closing this yield gap.

8.9.2   Certification and Standards

The process of obtaining organic certification can be costly and complex, particularly for small-scale farmers. Biodynamic farming is often practiced within the broader biodynamic agriculture framework, but some farmers choose to pursue biodynamic certification through organizations such as Demeter International. However, obtaining biodynamic certification can be complex and requires adherence to specific standards and practices. There is also the challenge of maintaining consistent standards across different countries.

8.9.3 Supply Chain and Logistics: Developing Efficient Supply Chains for Organic Products

Developing efficient supply chains for organic products can be challenging due to the smaller scale of operations and the need for segregation from nonorganic products to avoid contamination.

8.9.4 Knowledge and Technology Transfer

There is a need for more research into organic practices and better dissemination of knowledge and technology among organic farmers to improve efficiency and productivity.

8.10 Future Vista

The future of biodynamic agriculture looks promising but requires continued effort in terms of policy support, research, and education to overcome existing challenges. Despite its challenges, biodynamic agriculture continues to gain recognition and popularity among farmers, consumers, and policymakers who value its holistic and sustainable approach to food production. As concerns about climate change, soil degradation, and food quality intensify, biodynamic farming offers a viable alternative that prioritizes ecological health, social responsibility, and spiritual well-being. Innovations in biodynamic agriculture techniques and broader consumer acceptance can potentially make biodynamic agriculture more competitive and sustainable. As the sector grows, it will play a crucial role in addressing global issues like environmental sustainability, health, and food. Continued research, education, and support for biodynamic farming practices can help expand its adoption and contribute to a more resilient and regenerative agricultural system (Muhie 2022).

References

Clitheroe C (2019) Permaculture the documentary: how it started (documentary). DogsGoWoof Productions

Dervash MA, Bhat RA, Mushtaq N, Singh DV (2018) Dynamics and importance of soil Mesofauna. Int J Adv Res Sci Eng 7:2010–2019

Jackson G, Raster A, Shattuck W (2011) An analysis of the impacts of health insurance rebate initiatives on community supported agriculture in southern Wisconsin. J Agric Food Syst Communnity Dev 2:287–296. https://doi.org/10.5304/jafscd.2011.021.002

Muhie SH (2022) Novel approaches and practices to sustainable agriculture. J Agric Food Res 10:100446. https://doi.org/10.1016/j.jafr.2022.100446

Paull J (2011) Attending the first organic agriculture course: Rudolf Steiner's agriculture course at Koberwitz, 1924. Eur J Soc Sci 21:64–70

Rigolot C, Quantin M (2022) Biodynamic farming as a resource for sustainability transformations: potential and challenges. Agric Syst 200:103424. https://doi.org/10.1016/j.agsy.2022.103424

Willer H, Trávníček J, Meier C, Schlatter B (eds) (2022) The world of organic agriculture. Statistics and emerging trends. Research Institute of Organic Agriculture FiBL, Frick, and IFOAM—Organics International, Bonn

Yang X, Dai X, Zhang Y (2024) The government subsidy policies for organic agriculture based on evolutionary game theory. Sustain For 16:2246. https://doi.org/10.3390/su16062246

Zeng X, Gao H, Wang R, Majcher MB, Woon SJ, Wenda C, Eggleton P, Griffiths HM, Ashton LA (2024) Global contribution of invertebrates to forest litter decomposition. Ecol Lett 27:e14423. https://doi.org/10.1111/ele.14423

References

Chapter 9
Microbes and Ecosystem Cybernetics

Abstract The field of "ecosystem cybernetics" emerged as a scientific discipline with Norbert Wiener's articulation in 1948, further developed by influential figures such as Margalef (1968) and Odum (1971). Margalef conceptualized ecosystems as complex systems governed by feedback loops and species interactions, emphasizing their interconnectedness and emergent properties. Odum expanded upon these ideas, illustrating how feedback and control mechanisms regulate ecosystem stability. In this, it is detailed, how soil organisms play vital roles in ecosystem cybernetics, contributing to nutrient cycling, energy flow, and feedback mechanisms within terrestrial ecosystems. Besides, the challenges and future directions are also discussed.

9.1 Introduction

The "ecosystem cybernetics" was articulated as a scientific discipline by Norbert Wiener (1948). Afterwards, Margalef (1968) and Odum (1971) were influential figures in incorporating cybernetics into ecological theory. Margalef's conceptualization of ecosystems as complex systems governed by feedback loops and interactions among species laid the foundation for understanding the emergent properties of ecosystems. He emphasized the interconnectedness of various components within ecosystems and how their interactions give rise to macroscopic behaviors. Similarly, Odum expanded upon these ideas by illustrating how feedback and control mechanisms operate within ecosystems. He highlighted the self-regulating nature of ecosystems, wherein internal processes such as nutrient cycling and energy flow help maintain stability and balance over time. Odum's analogy of a thermostat-controlled furnace provided a clear illustration of how feedback regulation works to maintain a stable state in ecological systems.

Soil organisms play vital roles in ecosystem cybernetics, contributing to the regulation of nutrient cycling, energy flow, and feedback mechanisms within terrestrial ecosystems, which are conversed in this chapter. Besides, the challenges and future directions of ecosystem cybernetics are also discussed.

Soil organisms contribute to ecosystem cybernetics through the following ways (Collins 2024):

9.1.1   Nutrient Cycling

Soil organisms are crucial drivers of nutrient cycling processes in ecosystems. Microbes, such as bacteria and fungi, decompose organic matter, releasing nutrients like carbon, nitrogen, phosphorus, and sulfur into the soil. This decomposition process, along with activities of soil fauna like earthworms and arthropods, breaks down complex organic compounds into simpler forms that are readily available for plant uptake. Soil organisms also participate in processes like nitrogen fixation, nitrification, denitrification, and mineralization, which regulate nutrient availability and cycling in the ecosystem.

9.1.2   Energy Flow

Soil organisms contribute to energy flow within ecosystems by decomposing organic matter and releasing energy stored in organic compounds. Through metabolic activities such as respiration, fermentation, and predation, soil organisms transform organic materials into energy-rich molecules that fuel biological processes in the soil and support the growth and functioning of plants and other organisms within the ecosystem.

9.1.3   Feedback Mechanisms

Soil organisms are involved in feedback mechanisms that regulate ecosystem dynamics and stability. For example, microbial communities respond to changes in environmental conditions, such as moisture levels or nutrient availability, by adjusting their metabolic activities and population dynamics. These microbial responses can influence soil fertility, nutrient cycling rates, and the availability of resources for plant growth, thus shaping feedback loops that regulate ecosystem function.

9.1.4   Soil Structure and Stability

Soil organisms, particularly earthworms and soil-dwelling arthropods, play essential roles in soil structure formation and stability. Their burrowing activities aerate the soil, enhance water infiltration and drainage, and promote the aggregation of soil particles, leading to improved soil structure and fertility. Additionally, soil microbes produce extracellular substances like polysaccharides and glomalin, which bind soil particles together, contributing to soil stability and resilience against erosion and degradation.

Thus, soil organisms are integral components of ecosystem cybernetics, contributing to the regulation of nutrient cycling, energy flow, and feedback mechanisms that govern ecosystem dynamics and sustainability. Understanding the roles and interactions of soil organisms is essential for effective ecosystem management and conservation strategies aimed at maintaining soil health and ecosystem resilience.

9.2 Understanding Ecosystem Cybernetics

Ecosystem cybernetics draws from cybernetics, the science of communication and control systems in machines and living things. In ecosystems, this involves understanding how various components (biotic and abiotic) interact to regulate system behavior, such as nutrient cycling, energy flow, and response to disturbances. This approach helps in elucidating the complex feedback loops and interactions that maintain ecosystem stability and function (Romero 2020). Summarization of the key aspects of microbial roles and their implications in ecosystem cybernetics is simplified in Table 9.1.

9.3 Microbes and Ecosystem Cybernetics

Microbes and ecosystem cybernetics represent a fascinating intersection of biology and systems theory. In this context, ecosystem cybernetics refers to the study of regulatory mechanisms within ecological systems that maintain their health, stability, and functionality. Microbes, including bacteria, archaea, fungi, and viruses, are central to these regulatory processes, serving as both indicators and drivers of ecosystem health and resilience (Molefe et al. 2023).

9.4 Role of Microbes in Ecosystem Cybernetics

Microbes are fundamental to nearly every aspect of ecosystem, and the regulatory protocols in an ecological circuitry works in the following manner (Nica et al. 2023):

Table 9.1 Microbes and ecosystem cybernetics

Aspect	Description	References
Nutrient cycling	Microbes drive essential nutrient cycles such as carbon, nitrogen, and sulfur cycles, facilitating the decomposition of organic matter and nutrient recycling	Lawrence Berkeley National Laboratory (2024)
Environmental interactions	Microbial community structures are influenced by environmental factors like pH, salinity, and temperature, affecting their functional roles in ecosystems	Graham et al. (2016)
Microbial community dynamics	The composition and diversity of microbial communities are critical for ecosystem functioning, with core microbiomes contributing to stability and resilience	Pita et al. (2018)
Biogeochemical processes	Microbes mediate key biogeochemical processes such as methane oxidation, nitrogen fixation, and sulfate reduction, impacting global environmental systems	Graham et al. (2016)
Host-microbe interactions	In host-associated systems like the sponge holobiont, microbes contribute to host health and environmental adaptability through symbiotic relationships	Pita et al. (2018)
Microbial genomics	Advances in genomic technologies help in understanding the functional capabilities of microbial communities and their roles in ecosystem processes	Pita et al. (2018)
Predictive modeling	Integration of microbial community data with environmental variables enhances predictive models of ecosystem processes, aiding in better ecosystem management	Graham et al. (2016)

9.4.1 Nutrient Cycling

Microbes are key players in the biogeochemical cycles of carbon, nitrogen, phosphorus, and other essential nutrients. Through processes like nitrogen fixation, decomposition, and mineralization, they convert nutrients into forms usable by other organisms, promoting ecosystem productivity.

9.4.2 Primary Production

In aquatic systems, photosynthetic microbes such as cyanobacteria and microalgae contribute significantly to primary production, generating organic compounds through photosynthesis that form the base of aquatic food webs.

9.4.3 Soil Structure and Health

Soil microbes, including bacteria and fungi, help build soil structure and maintain its health. They are involved in the formation of soil aggregates and influence soil porosity and water retention, impacting plant growth and soil erosion.

9.4.4 Climate Regulation

Microbial activities influence the global climate system. For example, methane-producing archaea and carbon-sequestering cyanobacteria play roles in the greenhouse gas dynamics, affecting global warming and climate change.

9.4.5 Resistance and Resilience to Disturbances

Microbial diversity and function are crucial for ecosystem resistance to and recovery from disturbances, whether natural or anthropogenic. They can adapt to changes, continue their regulatory roles, and thus support ecosystem recovery.

9.4.6 Feedback Loops

Microbes are involved in negative and positive feedback loops that regulate ecosystem responses to environmental changes. For instance, soil microbes can either produce or consume greenhouse gases, influencing climate feedback mechanisms.

9.5 Research and Applications

Studying microbial roles in ecosystem cybernetics often involves techniques from molecular biology, bioinformatics, and ecological modeling. DNA sequencing and metagenomics allow researchers to analyze microbial communities and their functions without the need for culturing them in the lab. These insights can be integrated into ecological models to predict how ecosystems respond to various stressors (Nam et al. 2023).

Applications of this knowledge include (Nam et al. 2023) the following:

(i) *Bioremediation*: Utilizing microbes to degrade pollutants in contaminated environments.

(ii) *Agriculture*: Developing microbial inoculants to enhance soil fertility and crop resistance to pests and diseases.
(iii) *Climate Change Mitigation*: Harnessing microbial processes to sequester carbon or reduce methane emissions.

9.6   Challenges and Future Directions

One of the biggest challenges in integrating microbes into ecosystem cybernetics is the sheer complexity and variability of microbial communities across spatial and temporal scales. Moreover, predicting the behavior of these communities under changing environmental conditions remains a significant hurdle (Prosser and Martiny 2020).

Future research in microbial ecosystem cybernetics will likely focus on enhancing models with more accurate microbial data, exploring unknown microbial interactions, and applying this knowledge to manage ecosystems sustainably in the face of global environmental changes. This will include more sophisticated bioinformatics tools to handle large datasets and refined ecological models that better simulate the real-world complexities of microbial dynamics in ecosystems.

References

Collins LT (2024) CyberGaia: earth as cyborg. Humanit Soc Sci Commun 11:322. https://doi. org/10.1057/s41599-024-02822-y
Graham EB, Knelman JE, Schindlbacher A, Siciliano S, Breulmann M, Yannarell A, Beman JM, Abell G, Philippot L, Prosser J, Foulquier A, Yuste JC, Glanville HC, Jones DL, Angel R, Salminen J, Newton RJ, Bürgmann H, Ingram LJ, Hamer U, Siljanen HMP, Peltoniemi K, Potthast K, Bañeras L, Hartmann M, Banerjee S, Yu R-Q, Nogaro G, Richter A, Koranda M, Castle SC, Goberna M, Song B, Chatterjee A, Nunes OC, Lopes AR, Cao Y, Kaisermann A, Hallin S, Strickland MS, Garcia-Pausas J, Barba J, Kang H, Isobe K, Papaspyrou S, Pastorelli R, Lagomarsino A, Lindström ES, Basiliko N, Nemergut DR (2016) Microbes as engines of ecosystem function: when does community structure enhance predictions of ecosystem processes? Front Microbiol 7:214. https://doi.org/10.3389/fmicb.2016.00214
Lawrence Berkeley National Laboratory (2024) Microbes to ecosystems. Retrieved from Lawrence Berkeley National Laboratory. https://www.lbl.gov/research/microbes-to ecosystems/
Margalef R (1968) Perspectives in ecological theory. University of Chicago Press, Chicago
Molefe RR, Amoo AE, Babalola OO (2023) Communication between plant roots and the soil microbiome; involvement in plant growth and development. Symbiosis 90:231–239. https:// doi.org/10.1007/s13199-023-00941-9
Nam NN, Do HDK, Trinh KTL, Lee NY (2023) Metagenomics: an effective approach for exploring microbial diversity and functions. Food Secur 12:2140. https://doi.org/10.3390/foods12112140
Nica I, Chiriță N, Delcea C (2023) Towards a sustainable future: economic cybernetics in analyzing Romania's circular economy. Sustain For 15(19):14433. https://doi.org/10.3390/su151914433
Odum EP (1971) Fundamentals of ecology, 3rd edn. W. B. Saunders Co., Philadelphia
Pita L, Rix L, Slaby BM, Franke A, Hentschel U (2018) The sponge holobiont in a changing ocean: from microbes to ecosystems. Microbiome 6:46. https://doi.org/10.1186/s40168-018-0428-1

Prosser JI, Martiny JBH (2020) Conceptual challenges in microbial community ecology. Philos Trans R Soc Lond Ser B Biol Sci 375(1798):20190241. https://doi.org/10.1098/rstb.2019.0241

Romero A Jr (2020) Hypogean communities as cybernetic systems: implications for the evolution of cave biotas. Diversity 12:413. https://doi.org/10.3390/d12110413

Wiener N (1948) Cybernetics or control and communication in the animal and the machine. MIT Press, Cambridge

Chapter 10
Soil Remediation: *Biological Approaches, Regulatory Frameworks, and Circular Economy*

Abstract Soil pollution, stemming from industrial activities, agriculture, mining, waste disposal, and urbanization, poses significant threats to soil quality, biodiversity, and human health. Efforts to address soil pollution involve prevention, remediation, and sustainable land management practices. Biological approaches to soil remediation, such as bioremediation, harness the power of living organisms like plants, microorganisms, and fungi to mitigate pollution and restore environmental health. This eco-friendly technology offers a promising avenue for restoring contaminated soil by utilizing natural processes. For instance, hyperaccumulator plants excel at absorbing heavy metals, while microorganisms break down organic pollutants into harmless by-products through biodegradation. Fungi are particularly effective at degrading complex organic compounds and can thrive in diverse soil environments. This chapter will discuss soil remediation through biological procedures and the regulatory frameworks. Besides, the role of microbes (bioremediation) in circular economy in addressing soil pollution is also discussed.

10.1 Introduction

Soil pollution refers to the presence of harmful substances or contaminants in the soil that adversely affect its quality, fertility, and ability to support plant and animal life. These pollutants can originate from various sources, including industrial activities, agriculture, mining, waste disposal, and urbanization. The consequences of soil pollution are far-reaching, affecting not only soil quality but also water quality, air quality, and human health. Contaminated soils can lead to the accumulation of toxic substances in food crops, posing risks to human health through the ingestion of contaminated food and water (Nyiramigisha and Sajidan 2021). Soil pollution also threatens biodiversity, as it can disrupt soil microbial communities and negatively impact plant and animal species. Efforts to address soil pollution involve a combination of prevention, remediation, and sustainable land management practices. Strategies include improving industrial processes to minimize pollution, promoting sustainable agricultural practices such as organic farming and agroecology, implementing effective waste management systems, and employing remediation techniques such as bioremediation, phytoremediation, and mycoremediation.

Biological approaches to soil remediation leverage the power of living organisms, including plants, microorganisms, and fungi, to mitigate soil pollution and restore environmental health. Soil pollution is a pressing concern with far-reaching consequences, and biological remediation stands as a promising avenue for addressing it. By engaging the natural processes of various organisms, bioremediation offers a sustainable and effective means of restoring contaminated soil. This technique is versatile, adaptable to a range of pollutants and environmental conditions. For instance, certain plants, known as hyperaccumulators, excel at absorbing and concentrating heavy metals from the soil, effectively removing them from the environment. Microorganisms, including bacteria and fungi, play vital roles in breaking down organic pollutants like pesticides and petroleum hydrocarbons into harmless byproducts through processes such as biodegradation (Datta et al. 2020). Fungi, in particular, are adept at breaking down complex organic compounds and can thrive in diverse soil environments. Therefore, keeping the essence of this novel eco-friendly technology in mind, soil remediation through biological procedures and regulatory frameworks is discussed in this chapter. Besides, the role of microbes (bioremediation) in circular economy for highlighting the potential of this sustainable approach in addressing soil pollution is also discussed.

10.2 Bioremediation: Magnifying the Power of Hidden Beauties

Bioremediation: *Magnifying the Power of Hidden Beauties* beautifully encapsulates the essence of bioremediation as a process that harnesses the innate capabilities of microorganisms, plants, and fungi to restore contaminated environments. Just as a magnifying glass reveals the intricate details of hidden beauties, bioremediation uncovers the latent potential within nature to heal and rejuvenate polluted ecosystems. At its core, bioremediation is about recognizing the inherent beauty and resilience of the natural world, even in the face of environmental degradation. By amplifying the power of microbial communities, plant roots, and fungal networks, contaminated sites can be transformed into thriving habitats once again (Mir et al. 2022).

Through bioremediation, we delve beneath the surface of polluted soils and waters, uncovering the remarkable abilities of microorganisms to degrade toxic pollutants into harmless byproducts. Like hidden treasures waiting to be discovered, these microbial communities hold the key to unlocking the remediation potential of contaminated environments (Datta et al. 2020).

Bioremediation methods can be tailored to specific contaminants and site conditions, offering a targeted and efficient approach to soil remediation. Additionally, compared to traditional remediation methods, biological techniques often incur lower costs, minimize environmental disruption, and promote long-term sustainability. However, it's crucial to recognize that while bioremediation holds tremendous potential, it's not a panacea. Factors such as the type and concentration of pollutants, soil properties, climate, and presence of competing organisms can influence its effectiveness. In some cases, a combination of biological, chemical, and physical remediation methods may be necessary for comprehensive soil remediation (Parnian et al. 2022). Nonetheless, by harnessing the power of living organisms, biological remediation offers a path toward restoring soil health, preserving ecosystems, safeguarding human health, and promoting sustainable land management practices.

These methods are often more sustainable, cost-effective, and environmentally friendly compared to traditional remediation techniques. Some significant biological approaches to soil remediation are as follows:

10.3 Phytoremediation

Phytoremediation utilizes plants to extract, degrade, or immobilize contaminants from the soil. Plants can uptake contaminants through their roots and either store them in their tissues (phytoextraction), metabolize them within their tissues (phytodegradation), or facilitate their breakdown by associated microorganisms (rhizodegradation) (Ozturk et al. 2015a, b; Dervash et al. 2023). Examples of plants commonly used in phytoremediation include sunflowers for heavy metals, poplars for organic pollutants, and willows for wastewater treatment.

10.4 Microbial Bioremediation

Bioremediation employs microorganisms such as bacteria, fungi, and archaea to degrade or transform contaminants in the soil into less harmful substances. Microbial bioremediation can occur through various processes, including aerobic and anaerobic degradation, cometabolism, and bioaccumulation. Common microbial species used in bioremediation include *Pseudomonas*, *Bacillus*, *Rhizobium*, and *Mycobacterium* (Mir et al. 2022).

10.4.1 Bacterial Remediation

(a) *Streptomyces* species indeed play a crucial role in soil pollution abatement due to their unique characteristics and metabolic capabilities. One example of *Streptomyces* species commonly involved in soil remediation is *Streptomyces coelicolor* which is a filamentous soil bacterium known for its prolific production of secondary metabolites with antimicrobial and biodegradative properties. This species produces a diverse array of enzymes and bioactive compounds that enable it to degrade a wide range of organic pollutants, including hydrocarbons, aromatic compounds, pesticides, and industrial chemicals. In soil pollution abatement efforts, *Streptomyces coelicolor* and other *Streptomyces* species can be applied to contaminated sites to facilitate the degradation and detoxification of organic pollutants. These bacteria thrive in soil environments and can actively metabolize pollutants through enzymatic reactions and metabolic pathways, thereby reducing their concentration and environmental impact. Additionally, *Streptomyces* species are known for their ability to produce extracellular enzymes, such as ligninases and cellulases, which enable them to degrade recalcitrant organic compounds like polychlorinated biphenyls (PCBs) and polycyclic aromatic hydrocarbons (PAHs). This capability makes *Streptomyces* species valuable agents for soil remediation, particularly in environments contaminated with persistent organic pollutants (Khan et al. 2023).

(b) While *Rhizobium* species are primarily known for their role in nitrogen fixation through symbiotic associations with leguminous plants, some strains, such as *Rhizobium leguminosarum* and *Rhizobium etli*, have been found to exhibit bioremediation capabilities beyond nitrogen fixation. These strains have demonstrated the ability to degrade organic pollutants and immobilize heavy metals in soil and water environments. *Rhizobium*-mediated rhizoremediation involves harnessing the plant-microbe interactions in the rhizosphere to enhance microbial activity and pollutant degradation (Goyal and Habtewold 2023). This approach utilizes leguminous plants to stimulate the growth of *Rhizobium* strains and other rhizospheric microorganisms, creating a conducive environment for bioremediation processes. As the plants grow and develop root systems, they release exudates and organic compounds into the soil, which serve as energy sources for the rhizospheric microbial community. The microbial activity in the rhizosphere, facilitated by the presence of *Rhizobium* strains, promotes the degradation, immobilization, or transformation of organic pollutants and heavy metals, thereby contributing to soil and water remediation efforts. Rhizobium-mediated rhizoremediation offers a sustainable and environmentally friendly approach to pollutant cleanup, leveraging the natural interactions between plants and beneficial soil bacteria to enhance remediation outcomes (Fahde et al. 2023).

(c) *Mycobacterium* species are indeed notable for their ability to degrade various organic compounds, making them valuable contributors to bioremediation efforts. One prominent example of *Mycobacterium* strains used in bioremediation projects is *Mycobacterium vanbaalenii* strain PYR-1. *Mycobacterium vanbaalenii* strain PYR-1 has demonstrated significant capabilities in degrading a wide range of organic pollutants, particularly polycyclic aromatic hydrocarbons (PAHs). PAHs are persistent environmental pollutants often found in contaminated soil and water due to industrial activities, vehicle emissions, and improper waste disposal. *Mycobacterium vanbaalenii* PYR-1 is capable of metabolizing various PAHs, including high-molecular-weight compounds like benzo[a]pyrene, into simpler, less toxic compounds through aerobic degradation processes. This strain has been utilized in bioremediation projects to clean up soil and water contaminated with petroleum products, industrial chemicals, and hazardous waste containing PAHs. By harnessing the metabolic capabilities of *Mycobacterium vanbaalenii* PYR-1, bioremediation strategies aim to degrade PAHs and other organic pollutants, thereby reducing their environmental impact and restoring ecosystem health (Qutob et al. 2022).

(d) *Pseudomonas* species are indeed renowned for their metabolic versatility and their ability to degrade a diverse array of organic compounds, making them valuable assets in bioremediation efforts. One notable example of *Pseudomonas* strains commonly used in bioremediation projects is *Pseudomonas putida*. *Pseudomonas putida* is a well-studied bacterium known for its robust metabolic capabilities and its proficiency in degrading various organic pollutants. This strain possesses enzymes and metabolic pathways that enable it to metabolize hydrocarbons, aromatic compounds, pesticides, and other organic pollutants commonly found in contaminated soil and water environments. In bioremediation projects, *Pseudomonas putida* strains are often applied to remediate soil and water contaminated with petroleum products, industrial chemicals, and organic pollutants (Kivisaar 2020). These bacteria can break down complex organic molecules into simpler, less toxic compounds through aerobic degradation processes, thereby reducing the concentration and environmental impact of pollutants.

10.4.2   Mycoremediation

Mycoremediation utilizes fungi, particularly species of mushrooms, to degrade or sequester contaminants in the soil. Fungi produce enzymes that can break down complex organic molecules, making them effective for remediation of pollutants such as petroleum hydrocarbons, pesticides, and heavy metals (Akpasi et al. 2023). Some examples of fungi used in mycoremediation include oyster mushrooms (*Pleurotus ostreatus*) (Chun et al. 2019) and white-rot fungi (*Phanerochaete chrysosporium*) (Akpasi et al. 2023).

10.4.2.1   Mycorrhizal Fungi in Pollution Remediation

Mycorrhizal fungi form symbiotic relationships with plant roots, extending their hyphal networks into the soil far beyond the root zone. This relationship is beneficial for both the plants and fungi, as it enhances nutrient uptake for the plants and provides carbohydrates derived from photosynthesis to the fungi (Raina et al. 2020). In polluted environments, these associations can be particularly valuable in the following manner:

(i) *Enhanced Plant Growth and Survival*: Mycorrhizal fungi improve plant health by increasing nutrient and water uptake. Healthy plants are more resilient and can grow better in polluted environments, which can be crucial for establishing vegetation in degraded soils.

(ii) *Stabilization of Pollutants*: Some mycorrhizal fungi can immobilize heavy metals or other pollutants by binding them in the fungal biomass or by transforming them into less toxic forms. This process, known as "phytostabilization," helps to prevent the spread of pollutants through erosion, leaching, or runoff.

(iii) *Phytoremediation Support*: By improving plant health, mycorrhizal fungi aid phytoremediation efforts, where plants are used to extract, sequester, or degrade environmental contaminants. Fungi enhance the tolerance of host plants to high levels of pollution and stress, thus increasing the efficiency of phytoremediation (Ashraf et al. 2010; Bhat et al. 2017).

10.4.2.2 Saprotrophic and Endophytic Fungi in Pollution Remediation

In addition to mycorrhizal fungi, other fungal types also play roles in pollution remediation:

(a) *Saprotrophic Fungi*: These fungi decompose dead organic matter, releasing enzymes capable of breaking down various pollutants. For example, certain saprotrophic fungi can degrade polycyclic aromatic hydrocarbons (PAHs), pesticides, and industrial chemicals. Their ability to break down lignin and cellulose makes them particularly effective against complex organic pollutants.

(b) *Endophytic Fungi*: Living within plant tissues, endophytic fungi can enhance the host plant's ability to tolerate pollutants. They can also directly participate in the degradation of pollutants absorbed by the plants, thus contributing to the overall efficiency of phytoremediation strategies.

10.4.2.3 Research and Application Challenges

While the potential for fungi in pollution remediation is significant, there are several challenges and considerations for practical applications:

(a) *Site-Specific Conditions*: The effectiveness of fungal-assisted remediation depends heavily on environmental conditions such as soil pH, temperature, moisture, and the presence of other contaminants that might inhibit fungal growth or activity.

(b) *Selection of Fungal Species*: Identifying and utilizing the most effective fungal species for specific pollutants or environmental conditions are crucial. Not all fungi are equally capable of degrading all types of pollutants.

(c) *Scale and Control*: Scaling fungal-assisted remediation from laboratory or small-scale field trials to larger, more heterogeneous sites can be challenging.

Maintaining optimal conditions for fungal activity in diverse environmental settings requires careful management.

(d) *Ecological Impact*: Introducing non-native fungi into an ecosystem can have unintended consequences. It is important to consider the ecological balance and potential impacts on native species and soil health.

Overall, fungi offer promising solutions for environmental remediation, capable of enhancing the resilience of ecosystems and degrading or stabilizing pollutants through natural processes. Continued research and development are needed to optimize these biological interactions for practical and sustainable pollution management strategies.

10.5   Regulatory Frameworks and Policy Initiatives for Soil Remediation and Conservation

Regulatory frameworks and policy initiatives for soil remediation and conservation vary by country and region but generally aim to address soil contamination, erosion, and degradation while promoting sustainable land management practices (Katsir et al. 2024). The key aspects of regulatory frameworks and policy initiatives for soil remediation and conservation include the following:

10.5.1   Soil Contamination Regulations

(i) *Pollution Prevention Laws*: Many countries have laws and regulations aimed at preventing soil contamination from industrial activities, hazardous waste disposal, and agricultural practices. These laws often require businesses to implement pollution prevention measures and adhere to environmental standards to minimize soil contamination.

(ii) *Site Remediation Requirements*: Regulations may require responsible parties to investigate and remediate contaminated sites to protect human health and the environment. Remediation standards and cleanup criteria are established to guide remediation efforts and ensure that contaminated sites are properly restored.

(iii) *Liability and Enforcement*: Regulatory frameworks often include provisions for assigning liability for soil contamination and enforcing compliance with remediation requirements. Polluters may be held financially responsible for cleanup costs, and regulatory agencies may impose penalties for noncompliance.

10.5.2   Soil Conservation Policies

(a) *Conservation Programs*: Governments may implement conservation programs to incentivize farmers and landowners to adopt soil conservation practices such as conservation tillage, cover cropping, and erosion control measures. These programs may provide financial incentives, technical assistance, and education to promote sustainable land management.

(b) *Soil Health Initiatives*: Soil health initiatives focus on promoting soil conservation and improving soil quality through measures such as soil testing, nutrient management planning, and biodynamic agriculture practices. These initiatives aim to enhance soil fertility, water retention, and resilience to erosion and degradation.

(c) *Land-Use Planning*: Land-use planning and zoning regulations may incorporate soil conservation considerations to guide development and land-use decisions. Conservation easements, buffer zones, and green infrastructure requirements are examples of planning tools used to protect soil resources and minimize environmental impacts.

10.5.3   International Agreements and Initiatives

(i) *Global Soil Partnership:* The Global Soil Partnership (GSP), established by the Food and Agriculture Organization (FAO) of the United Nations, aims to promote sustainable soil management and conservation worldwide. The GSP facilitates collaboration among governments, organizations, and stakeholders to address soil-related challenges and implement soil conservation initiatives (Erdogan et al. 2021).

(ii) *The United Nations Sustainable Development Goals*: The United Nations Sustainable Development Goals (SDGs) include targets related to soil conservation, land degradation neutrality, and sustainable agriculture. SDG 15 specifically addresses the importance of protecting, restoring, and promoting sustainable use of terrestrial ecosystems, including soil resources (Erdogan et al. 2021).

10.5.4   Research and Monitoring

(i) *Soil Monitoring Programs*: Governments may establish soil monitoring programs to assess soil quality, identify trends in soil degradation, and inform soil conservation efforts. Monitoring data helps policymakers, researchers, and land managers understand the status of soil resources and prioritize conservation actions.

(ii) *Research Funding and Collaboration*: Governments, research institutions, and international organizations invest in research and innovation to develop new soil remediation technologies, conservation practices, and sustainable land management strategies. Collaboration among researchers, policymakers, and stakeholders facilitates knowledge sharing and capacity building in soil conservation.

In this way, regulatory frameworks and policy initiatives for soil remediation and conservation play a critical role in protecting soil resources, promoting sustainable land management, and safeguarding environmental quality for current and future generations. Effective implementation and enforcement of these policies require collaboration among governments, stakeholders, and the public to address soil-related challenges and achieve environmental sustainability goals.

10.6 Bioremediation and the Circular Economy

Bioremediation and the circular economy are two concepts that can complement each other synergistically, contributing to more sustainable approaches to environmental management and resource utilization.

Bioremediation involves using living organisms to degrade, detoxify, or remove contaminants from the environment, such as soil, water, or air. By harnessing the natural abilities of microorganisms, plants, and fungi, bioremediation offers a cost-effective and environmentally friendly method for cleaning up polluted sites.

The circular economy, on the other hand, is an economic model that aims to minimize waste and resource consumption by maximizing the reuse, recycling, and regeneration of materials and products. Instead of the traditional linear "take-make-dispose" approach, the circular economy seeks to close the loop, keeping resources in use for as long as possible and extracting maximum value from them (Topliceanu et al. 2023; Tiwari et al. 2023).

Bioremediation fits well within the principles of the circular economy for several reasons:

(a) *Resource Efficiency*: Bioremediation often relies on natural processes and renewable resources, such as sunlight, water, and organic matter, to facilitate the degradation of contaminants. By using these resources more efficiently, bioremediation helps reduce the consumption of energy and raw materials associated with traditional remediation methods.

(b) *Waste Reduction*: Bioremediation can transform harmful pollutants into harmless byproducts or convert them into valuable resources. For example, some microorganisms can metabolize organic pollutants and convert them into carbon dioxide, water, and biomass. By converting waste into useful products, bioremediation contributes to the circular economy's goal of minimizing waste generation.

(c) *Closed-Loop Systems*: In some cases, bioremediation can be integrated into closed-loop systems where the byproducts of remediation are reused or recycled within the same ecosystem. For instance, plants used in phytoremediation may accumulate contaminants in their tissues, which can then be harvested and processed to recover valuable metals or nutrients.

(d) *Ecosystem Restoration*: Bioremediation not only cleans up polluted environments but also restores ecosystems and enhances biodiversity. By promoting the recovery of natural habitats and ecosystem services, bioremediation aligns with the circular economy's aim of regenerating natural capital and preserving ecosystem integrity.

In the current scenario of environmental upheaval, integrating bioremediation practices into the circular economy framework can help optimize resource use, minimize environmental impacts, and create more resilient and sustainable systems for managing contaminated sites and preserving environmental quality (Tiwari et al. 2023). By leveraging the power of nature to clean up pollution and regenerate ecosystems, we can move toward a more circular and sustainable future.

References

Akpasi SO, Anekwe IMS, Tetteh EK, Amune UO, Shoyiga HO, Mahlangu TP, Kiambi SL (2023) Mycoremediation as a potentially promising technology: current status and prospects—a review. Appl Sci 13(8):4978. https://doi.org/10.3390/app13084978

Ashraf M, Ozturk M, Ahmad MSA (2010) Toxins and their phytoremediation. In: Plant adaptation and phytoremediation, pp 1–32

Bhat RA, Dervash MA, Mehmood MA, Bhat MS, Rashid A, Bhat JIA, Singh DV, Lone R (2017) Mycorrhizae: a sustainable industry for plantand soil environment. In: Varma A et al (eds) Mycorrhiza—nutrient uptake, biocontrol, ecorestoration. Springer International Publishing, pp 473–502

Chun SC, Muthu M, Hasan N, Tasneem S, Gopal J (2019) Mycoremediation of PCBs by *Pleurotus ostreatus*: possibilities and prospects. Appl Sci 9:4185. https://doi.org/10.3390/app9194185

Datta S, Rajnish KN, Samuel MS, Pugazlendhi A, Selvarajan E (2020) Metagenomic applications in microbial diversity, bioremediation, pollution monitoring, enzyme and drug discovery. A review. Environ Chem Lett 18:1229–1241

Dervash MA, Yousuf A, Ozturk M, Bhat RA (2023) Phytoremediation of nuisance pollution. In: Phytosequestration: strategies for mitigation of aerial carbon dioxide and aquatic nutrient pollution. Springer International Publishing, Cham, pp 89–93

Erdogan HE, Havlicek E, Dazzi C, Montanarella L, Liedekerke MV, Vrščaj B, Krasilnikov P, Khasankhanova G, Vargas R (2021) Soil conservation and sustainable development goals (SDGs) achievement in Europe and Central Asia: which role for the European soil partnership? Int Soil Water Conserv Res 9:360–369. https://doi.org/10.1016/j.iswcr.2021.02.003

Fahde S, Boughribil S, Sijilmassi B, Amri A (2023) Rhizobia: a promising source of plant growth-promoting molecules and their non-legume interactions: examining applications and mechanisms. Agriculture 13:1279. https://doi.org/10.3390/agriculture13071279

Goyal RK, Habtewold JZ (2023) Evaluation of legume–rhizobial symbiotic interactions beyond nitrogen fixation that help the host survival and diversification in hostile environments. Microorganisms 11:1454. https://doi.org/10.3390/microorganisms11061454

Katsir S, Biswas AK, Urs K, Lenka NK, Jha P, Arora K (2024) Governing soils sustainably in India: establishing policies and implementing strategies through local governance. Soil Secur 14:100132. https://doi.org/10.1016/j.soisec.2024.100132

Khan S, Srivastava S, Karnwal A, Malik T (2023) *Streptomyces* as a promising biological control agents for plant pathogens. Front Microbiol 14:1285543. https://doi.org/10.3389/fmicb.2023.1285543

Kivisaar M (2020) Narrative of a versatile and adept species *Pseudomonas putida*. J Med Microbiol 69:324–338. https://doi.org/10.1099/jmm.0.001137

Mir S, Dervash MA, Shikari AB, Showket W (2022) Microbial consortium: a biotechnological tool for enhanced bioremediation in pollution effected environment. In: Environmental biotechnology: sustainable remediation of contamination in different environs, CRC Press. edited by Bhat RA, Dervash MA, Hakeem KR, Masoodi KZ

Nyiramigisha P, Sajidan K (2021) Harmful impacts of heavy metal contamination in the soil and crops grown around dumpsites. Rev Agric Sci 9:271–282. https://doi.org/10.7831/ras.9.0_271

Ozturk M, Ashraf M, Aksoy A, Ahmad MSA (eds) (2015a) Phytoremediation for green energy. Springer Science+Business Media, NY

Ozturk M, Ashraf M, Aksoy A, Ahmad MSA, Hakeem KR (eds) (2015b) Plants, pollutants and remediation. Springer, Amsterdam

Parnian A, Pirasteh-Anosheh H, Ozturk M, Unal D, Yilmaz DD, Altay V (2022) Bioremediation. Mai J Sci 49:1–9

Qutob M, Rafatullah M, Muhammad SA, Alosaimi AM, Alorfi HS, Hussein MA (2022) A review of pyrene bioremediation using *mycobacterium* strains in a different matrix. Fermentation 8:260. https://doi.org/10.3390/fermentation8060260

Raina S, Yahmed NB, Bhat RA, Dervash MA (2020) Mycoremediation: a sustainable tool for abating environmental pollution. In: Hakeem KR, Bhat RA, Qadri H (eds) Bioremediation and biotechnology: sustainable approaches to pollution degradation. Springer Nature, pp 269–292. https://doi.org/10.1007/978-3-030-35691-0_13

Tiwari S, Mohammed KS, Mentel G, Majewski S, Shahzadi I (2023) Role of circular economy, energy transition, environmental policy stringency, and supply chain pressure on CO_2 emissions in emerging economies. Geosci Front 15:101682. https://doi.org/10.1016/j.gsf.2023.101682

Topliceanu L, Puiu PG, Drob C, Topliceanu VV (2023) Analysis regarding the implementation of the circular economy in Romania. Sustain For 15:333. https://doi.org/10.3390/su15010333

Index

© The Author(s), under exclusive license to Springer Nature Switzerland AG 2024   119
M. A. Dervash et al., *Soil Organisms*, SpringerBriefs in Microbiology,
https://doi.org/10.1007/978-3-031-66293-5